Articulatory Speech Synthesis from the Fluid Dynamics of the Vocal Apparatus

Synthesis Lectures on Speech and Audio Processing

Editor
B.H. Juang, *Georgia Tech*

Articulation and Intelligibility
Jont B. Allen
2005

Articulatory Speech Synthesis from the Fluid Dynamics of the Vocal Apparatus

Stephen Levinson, Don Davis, Scot Slimon, and Jun Huang

ISBN: 78-3-031-01435-2   paperback
ISBN: 78-3-031-02563-1   ebook

DOI 10.1007/978-3-031-02563-1

A Publication in the Springer series
SYNTHESIS LECTURES ON SPEECH AND AUDIO PROCESSING
Lecture #9
Series Editor: B.H. Juang, *Georgia Tech*
Series ISSN
Synthesis Lectures on Speech and Audio Processing
Print 1932-121X   Electronic 1932-1678

Articulatory Speech Synthesis from the Fluid Dynamics of the Vocal Apparatus

Stephen Levinson
University of Illinois at Urbana Champaign

Don Davis
GD/Electric Boat

Scot Slimon
GD/Electric Boat

Jun Huang
SoundHound

SYNTHESIS LECTURES ON SPEECH AND AUDIO PROCESSING #9

ABSTRACT

This book addresses the problem of articulatory speech synthesis based on computed vocal tract geometries and the basic physics of sound production in it. Unlike conventional methods based on analysis/synthesis using the well-known source filter model, which assumes the independence of the excitation and filter, we treat the entire vocal apparatus as one mechanical system that produces sound by means of fluid dynamics. The vocal apparatus is represented as a three-dimensional time-varying mechanism and the sound propagation inside it is due to the non-planar propagation of acoustic waves through a viscous, compressible fluid described by the Navier-Stokes equations.

We propose a combined minimum energy and minimum jerk criterion to compute the dynamics of the vocal tract during articulation. Theoretical error bounds and experimental results show that this method obtains a close match to the phonetic target positions while avoiding abrupt changes in the articulatory trajectory. The vocal folds are set into aerodynamic oscillation by the flow of air from the lungs. The modulated air stream then excites the moving vocal tract. This method shows strong evidence for source-filter interaction.

Based on our results, we propose that the articulatory speech production model has the potential to synthesize speech and provide a compact parameterization of the speech signal that can be useful in a wide variety of speech signal processing problems.

KEYWORDS

articulatory speech synthesis, Navier-Stokes equations, computational fluid dynamics, human vocal apparatus, articulatory dynamics

Contents

Preface

The research reported in this book is but one part of a long history of research conducted at Bell Laboratories devoted to understanding the nature of the unique acoustic signal commonly called speech. Often, such research is presented in the form of methods of speech synthesis but it has far broader implications. One form of speech synthesis is that of synthesis by analysis in which natural speech is decomposed into a spectral representation of a small inventory of fundamental sounds that can be reordered and recombined to produce any desired spoken utterance. In fact, most commercially available speech synthesis technology is of this kind. An alternative approach, of which our work is an example, is an ab initio method in which a speech signal is generated based only on the physical acoustics and known geometry of the human vocal apparatus without recourse to any measurements of a natural speech signal. Often referred to as articulatory synthesis, this method produces synthetic speech of inferior quality to that of analysis/synthesis but has the potential to provide deeper understanding of speech. One of the earliest examples of articulatory synthesis is the VODER of H. Dudley. A complete speaking machine operated by a skilled technician working at a mechanical keyboard, it was exhibited at the 1939 World's Fair to the delight of the public who heard it speak. The mechanism was a realization of the source-filter model upon which most of modern speech-processing technology is based. The model characterizes the vocal apparatus as a filter of time-varying frequency response. The airflow into the vocal tract is identified with the input to the filter whose output is heard as a speech signal. Several different implementations of the VODER followed. C. Coker and N. Umeda produced an elaborate computer simulation of the entire vocal apparatus including a video display of the vocal tract synchronized with the speech it produced. The speech signal was obtained by computing the eigenvalues of the Webster Equation for the time-varying vocal tract shape. It is worth noting that to accomplish this simulation in real time required the construction of a unique general purpose computer with a fast (for its day) cycle time. Subsequently, M. Portnoff showed how to obtain the speech signal from a complete solution to the Webster Equation and associated boundary conditions. Then, S. Levinson and C. Schmidt used this method to solve the vocal tract inverse problem, the determination of the vocal tract shape from the measured synthetic speech signal. The most elaborate simulation of this type is due to J. L. Flanagan and K. Ishizaka based on a nonlinear transmission-line form of the vocal tract filter. In all of these "first principle" approaches, certain simplifying assumptions were made to make the computation tractable. Among them were source-filter separation, the production of plane acoustic waves propagating independent of the air flow through a straight pipe approximation of the irregular curved vocal tract. The pipe was assumed to have a temporally and spatially varying circular cross section. Our experiments were intended to relax these assumptions by including the fluid dynamics inside a three-dimensional vocal tract with source-filter coupling. This required solving the notoriously difficult Navier-Stokes equations with time-varying boundary conditions. We were fortunate to have available a proprietary computer code developed for other purposes by D. Davis, S. Slimon, and their colleagues at the Electric Boat Corporation. Based on the well-known RANS approximation, it allowed for moving grids to account for time and space variation of the solutions resulting from the articulatory dynamics and source-filter interactions. The computed solutions yielded both intelligible speech and strong evidence of source filter coupling even with an aerodynamic oscillator model of vocal cord vibration. Finally, we are obliged to give a special mention of the role of J. L. Flanagan in our research and indeed in most of the research cited throughout this book. In his capacity as head of the Acoustics Research Department and later as director of the Information Principles Research Laboratory at Bell Laboratories, Dr. Flanagan either supervised or directly participated in all of the cited research on articulatory synthesis. Dr. Flanagan was convinced that results obtained from the study of speech synthesis could be used to advance all speech processing technologies. He often remarked that that speech synthesis from "first principles" would lead to the development of a parametric representation of the speech signal that would capture the information bearing features of the signal and would thus serve as a parsimonious basis for automatic speech recognition and speech compression (coding). Although the research reported herein does not satisfy Dr. Flanagan's goals, it is a step toward that end. We were greatly aided and encouraged by Dr. Flanagan's efforts including his arrangement of DARPA and NSF funding. We are most appreciative of the opportunity to contribute to his vision.

Stephen Levinson, Don Davis, Scot Slimon, and Jun Huang
July 2012

CHAPTER 1

Introduction

1.1 HISTORY OF SPEECH SYNTHESIS

Speech is the most important form of human communication. It consists of a sequence of sounds which are generated by the vocal apparatus and used as a vital tool for interaction among human beings. People have long been fascinated by the possibility of enabling inanimate machines with the power of speech, or at least something resembling it.

Early attempts to construct talking machines can be traced back to the eighteenth century. Interest in speech synthesis began during the Greek and Roman civilizations when clever deception gave voice to inanimate statues and gods. In 1779, Kratzenstein constructed a set of acoustic resonators which, when activated by a vibrating reed, produced a reasonable imitation of the steady-state vowels [1]. Twelve years later, Von Kempelen built a more elaborate machine that could generate connected utterances of speech. The Von Kempelen speaking machine used a bellows to supply air through a reed acting as the vocal cords. Voiced sounds, consonants, and nasals could be generated by exciting a resonator whose shape was determined by the operator's one hand and controlled by the fingers of the other hand. An elaborate version of this machine was later built and demonstrated by Sir Charles Wheatstone in 1879. The influence of Von Kempelen's machine also motivated Alexander Graham Bell in the late eighteenth century to model many of the articulators using a cast from a human skull and modelling the interior parts of the vocal tract in guttapercha. Interestingly enough, Bell and his brother were able to imitate many of the vowels and nasals, and their creative work was later acknowledged as a U.S. patent, dated 1876. The construction of mechanical machines to simulate voice sounds was not an easy task due to the numerous variations in pitch, intonation, etc., and the complex movements of the vocal apparatus. As an article in *Scientific American* published in 1871 [2] puts it, "Machines which, with more or less success, imitate human speech, are the most difficult, to construct, so many are the agencies engaged in uttering even a single word - lungs, larynx, tongue, palate, teeth, lips - so many are the inflections and variations of tone and articulation, that the mechanician finds his ingenuity taxed to the uttermost to imitate them."

The first electrical synthesizer which attempted to generate connected utterances was the Voder developed by Dudley et al [3] in 1939. The device was driven manually and consisted of ten parallel bandpass filters that spanned the entire frequency range of speech, and a set of resonances and antiresonances were applied to the excitation source. The resonance control box was excited by either a noise source or a buzz oscillator. The Voder achieved enormous success in producing reasonably intelligible speech quality and was later demonstrated at the world's fairs of 1939 in New York and 1940 in San Francisco. The evolution of electronic technology and the greater interest in human-machine communications sprang new and exciting dimensions in speech synthesis. The first was the development of the Vocoder machine by Dudley [4] based on the principles of speech analysis and synthesis. With the recent growth in high speed digital computers and the progress encountered in the development of integrated circuitry, analysis-synthesis techniques began to play an important role in speech communication to reduce the transmission bandwidth used for speech signals. The application of analysis-synthesis techniques extends to many areas where memory storage is limited. Synthetic speech in devices for human-machine communication has proved successful in producing reasonable human utterances and reducing message storage space significantly.

As an alternative to analysis-synthesis methods, speech can be synthesized according to certain linguistic and acoustic rules which convert a string of discrete symbols into speech. This is known as *text-to-speech* (TTS), or synthesis by rule. Prosodic information such as duration, pitch, and stress is necessary to describe the manner and strategy by which the utterance is to be spoken. The history of speech synthesis by rule extends back to the late 1950s when acoustic signals started to be recognized as individual speech sounds [5]. However, the first automatic TTS system was generated in 1961 by Kelley and Gerstman [6] which enabled direct transformation of linguistic units, supplemented with pitch and timing information, into segments of speech. Their rules for formant synthesis were later elaborated for British English by Holmes et al. in 1964 [7].

1.2   SPEECH PRODUCTION

In the previous section, a brief history of speech synthesis was outlined. Any kind of speech synthesis method is based on a certain type of speech production model. These models differ in their complexity and in the methods employed for achieving natural quality synthetic speech. Speech can be described as a sequence of sounds which are generated when a flow of air is disrupted or perturbed by the vocal apparatus, namely, lips, jaw, tongue, velum, and larynx. These are also known as the vocal articulators. A simplified diagram of the human vocal system for speech production is shown in Figure 1.1. The vocal tract is considered a nonuniform acoustic tube which extends from the vocal cords to the lips. The vocal cords are essentially shelf-like ligament-covered muscles embedded in the edges of the larynx. The region between the larynx and the velum is referred to as the *pharyngeal tract* and that between the velum and the lips as the *oral tract*. The cross-sectional area of the vocal tract varies from 0 to 20 cm^2 depending on the position of the articulators, which also determine the overall length of the vocal tract. The vocal tract length varies between 15 cm and 20 cm for different speakers and may change according to the sound produced. For an average adult male, the vocal tract is considered to be about 17 cm long in its rest position. For the production of nasal sounds, the nasal tract is acoustically coupled to the vocal tract by lowering the velum and forming a constriction at some point along the oral cavity. Under these conditions, the air flows through the pharynx into the nasal cavity and sound is radiated at the nostrils. For an average adult male, the overall length of the nasal tract is about 12 cm and its cross-sectional area changes mainly in the region close to the velum.

Speech sounds can be roughly classified into three categories, namely, *sonorants*, *fricatives*, and *plosives*, depending on their primary methods of excitation. *Voiced* sounds, including all sonorants

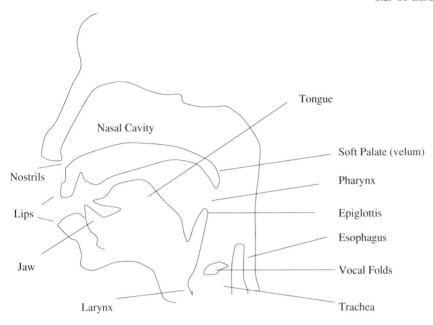

Tongue

Nasal Cavity

Soft Palate (velum)

Nostrils

Pharynx

Lips

Epiglottis

Esophagus

Jaw

Vocal Folds

Larynx

Trachea

Figure 1.1: A simplified diagram of the human vocal system.

and some fricatives and stops, are produced if the air passing through the larynx causes vibration of the vocal folds. Consequently, quasi-periodic pulses of air are produced whose period is determined by the mass and tension of the cords, as well as the subglottal pressure. Examples of voiced sounds are /i/ (as in *beet*), /n/ (as in *noise*), and /w/ (as in *where*). *Fricative* sounds are generated by forcing air through a constriction formed at some point in the vocal tract which results in a turbulent flow of air in that region. Constrictions can be labio-dental, dental, alveolar, palatal, or glottal. For example, the sound /θ/ (as in *thick*) is produced by a constriction in the dental region. The third class of sounds is referred to as *plosive*. These are produced when a build-up of pressure behind a complete closure in the vocal tract is suddenly released by the articulators. Closures can be labial, alveolar, palatal, or velar. A typical example is the sound /p/ (as in *pick*), produced when a closure in the labial region is formed. All the above speech sounds are considered as basic linguistic units and are often referred to as *phonemes*. Details of the physiology of speech production and a phonetic study of English sounds are described in great detail by Fant [8].

Currently, the most widely used speech production model assumes that the excitation source (vocal cords) and the "filter" (vocal tract) are separable; the vocal tract can be approximated as a series of abutting lossless acoustic tubes and the sound propagation inside the vocal tract can be modeled as a one-dimensional plane wave. These assumptions lead to the source-filter speech production which can be formulated mathematically as the linear prediction coding (LPC) model. Despite its simplification of the complicated speech production mechanism, the LPC model has resulted in tremendous success in a wide range of speech processing scenarios, such as feature extraction in speech recognition, speech production modelling in speech synthesis, and in speech coding. The details of the source-filter model and LP analysis will be presented in the next chapter.

1.3   CONTRIBUTIONS

During the last decades, progress in computational power, pattern recognition, signal processing, statistical modelling and other areas has opened a new phase in the area of speech signal processing. For example, the concatenative speech synthesis technique can synthesize highly intelligible speech sounds. For some specific domain applications, it can synthesize almost natural sounding speech by careful concatenation of word strings in a large database. However, there still remain specific setbacks that prevent current systems from achieving the goal of completely human-sounding speech. Furthermore, even if the current systems might have a lot of commercial success, they cannot help us understand the basics and some unsolved, yet important problems in human speech and language.

1.4   ORGANIZATION OF THE BOOK

This book is organized as follows. In Chapter 2, we review the relevant literature on speech synthesis and speech production modelling. The algorithms of estimation of dynamic articulatory parameters and the error analysis are developed in Chapter 3. In Chapter 4, we present the articulatory models. In Chapter 5, we present the excitational models. We discuss speech synthesis in Chapter 6. Finally, a conclusion and comments about some interesting future research topics are given in Chapter 7.

CHAPTER 2

Literature Review

2.1 OVERVIEW OF SPEECH SYNTHESIS TECHNIQUES

Speech synthesis technology can been classified into three broad categories: concatenative synthesis, formant synthesis, and articulatory synthesis.

Concatenative synthesis is currently the most popular technique in commercial and research text-to-speech (TTS) systems. This method relies on extracting model parameters from speech data (or often just storing raw waveforms) and concatenating these to create new utterances. There are several classes of speech models that allow varying degrees of control over the different characteristics of the voice quality: linear predictive coding (LPC), based on a source-filter model using a synthetic source; pitch-synchronous time-domain models; and sinusoidal models that represent the speech waveform as a sum of time-varying sinusoidal waves. Concatenative synthesis holds the promise of representing ill-understood fine details in the voice and replacing hand-optimization of rules with a data-driven approach to waveform generation. The intelligibility of the concatenative speech synthesis systems is good. However, they are not flexible in producing characterized sounds of different speakers and the naturalness of the synthesized speech is not satisfactory.

Formant synthesis is a parametric approach which applies a set of rules for controlling the frequencies and amplitudes of the formants and the characteristics of the excitation source. Although certain isolated phonetic units can be characterized almost solely by their formant frequencies and motions, the formant locations in continuous natural speech are heavily influenced by context. Because of this fact, the rules necessary to control a formant synthesizer are rather complex. Another feature of parametric formant synthesis is that the input parameter list can grow to be fairly large. A Klatt synthesizer can have more than 40 input parameters that need to be tracked over time. The shortcomings of the formant speech model are the fundamental limit to the naturalness of speech produced by rule, and the large number of parameters that need to be tracked.

Articulatory synthesis attempts to produce speech by first understanding how the vocal apparatus changes shape during speech production, then understanding the acoustic problem of how those movements translate into sounds. The input parameters to an articulatory synthesizer used to create speech include the positions of the model articulators, such as the tongue body height, and these parameters are specified as trajectories through time. Physical models of aerodynamic processes and of wave propagation inside a tube are used to convert the input parameters of the synthesizer into sound. Articulatory synthesis is potentially the most efficient way to generate speech waveforms with natural sounding, customized voices. However, many important problems need to be solved before an articulatory synthesizer can be used to produce better quality speech. One of the major impediments to the use of articulatory synthesis in creating natural sounding speech has been a lack of knowledge of the articulatory movement patterns. Second, a good vocal fold model needs to be coupled with the fluid model inside the vocal apparatus to generate the excitation signals for high quality speech synthesis. Third, there is a problem in identifying the articulatory degrees of freedom which are most salient to the production and propagation of sound. It is necessary to know the salient components of articulation because there are simply too many degrees of freedom to hope to make a practical articulatory synthesizer without such knowledge. Fourth, it is well-known that the sound propagation inside the vocal tract is a three-dimensional non-plane-wave propagation inside a viscous fluid described by the governing Navier-Stokes equation. How can we investigate the energy exchange between the convective and propagative components of the fluid flow and the effect of nonlinear fluid flow on the speech production? Finally, although articulatory synthesis is efficient in terms of the number of controlled parameters, the computational cost is very high. A new computational model which can capture the nonlinear effects in speech production with reasonable computational complexity is very important for the practical usage of articulatory synthesis.

2.1.1   CONCATENATIVE SYNTHESIS

A widely used method for converting a string of phonemes into an acoustic signal is the concatenation of segments (such as diphones) of naturally spoken utterances. In concatenative synthesis, segments of speech are excised from spoken utterances and are connected to form the desired speech signal. Currently, concatenative synthesis is the most commercially successful technique for speech generation. In December 2000, SpeechWorks announced: "Speechify 1.0, a product of the strategic partnership between SpeechWorks and AT&T, is the first to capitalize on 30 years of the AT&T Labs research developing human sounding, synthesized speech with sophisticated language analysis capability. With Speechify1.0, weather updates, traffic reports, search engine results, email, stock quotes and a wide range of other kinds of dynamic information from databases or websites can be accessed in real-time and easily understood by callers. The Speechify engine retrieves information from a database, synthesizes the text and outputs it in audio format over any phone."

There are two main problems in concatenative synthesis: *unit selection* and *speech model*. *Unit selection* involves defining the inventory of units as well as selecting the appropriate unit for a given phonetic (and prosodic) context. *Speech models* in speech synthesis include linear predictive coding (LPC), pitch-synchronous time-domain models such as pitch-period-sized waveform samples (PSOLA), and sinusoidal models that represent the speech waveform as a sum of time-varying sine waves. Although each of these models has its own merit, none has yet proven to satisfy all the desired properties of a TTS waveform model.

In this section, we will discuss the AT&T NextGen speech synthesis system, which is a typical example of the state-of-art concatenative synthesis system. Figure 2.1 shows the system diagram of the AT&T Next-Gen TTS system. The text normalization, linguistic processing such as syntactic analysis, word pronunciation, prosodic prediction, and prosody generation (translation between a symbolic representation to numerical values of fundamental frequency F_0, duration, and amplitude) is done by a Flextalk object that borrows heavily from AT&T Bell Labs' previous TTS system [9]. In the following, we will discuss in some detail the unit selection and synthesis back-end of the AT&T NextGen TTS system.

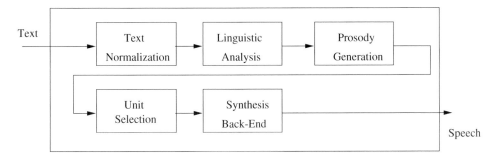

Figure 2.1: System architecture of the AT&T Next-Gen TTS.

Diphone synthesis has been popular for a number of years, due to the high intelligibility that such systems provide. They have the ability to preserve some of the coarticulation effects that are present at phoneme boundaries. However, such systems are handicapped by having a large number of distance units (in the range of 1000 - 3000 diphones, depending on language and phone-set chosen), for which it is not easy to create a sufficiently large database to capture all relevant coarticulation effects. Statistics become even worse if we demand that all relevant prosodic variations are covered.

The online unit selection of AT&T's NextGen TTS was adopted from ATR's CHATR system [10], [11], [12]. The CHATR system requires a set of speech units that can be classified into a small number of categories such that sufficient examples of each unit are available to make statistical selection viable. Hence, the original CHATR system uses phonemes as units. To avoid problems of concatenation at phoneme boundaries, a flexible joint technique is employed that allows moving unit boundaries. In order to arrive at a robust paradigm, the AT&T's NextGen TTS has chosen to use half phones as the basic units of synthesis in a way that allows synthesis from units ranging in size from diphones and phones to whole words and even phrases.

Most of the concatenative TTS systems use the statistical modelling technique from speech recognition fields to dynamically search for the sequence of synthesis units that minimizes context mismatch and concatenation costs. In this way, units with size varying from a fraction of a phone to many words can be used together in a dynamic way. Figure 2.2 shows the Viterbi search based on an inventory of multiple instances of each half-tone needed for synthesizing silence-/t/-/uw/-silence (the word "two"). More details can be found in [13].

In the context of speech synthesis based on concatenation of acoustic units, speech signals may be encoded by speech models. These models are required to ensure that the concatenation of selected acoustic units results in a smoothed transition from one acoustic unit to the next. One option in AT&T's NextGen TTS is the Harmonic plus Noise model (HNM) proposed by Stylianou [14]. HNM has the capability of effective waveform modifications such as controls of prosody and emo-

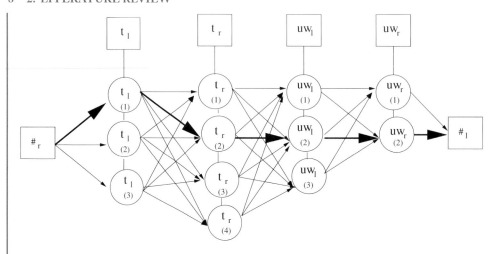

Figure 2.2: Viterbi search based on an inventory of multiple instances of each half-tone needed for synthesizing silence-/t/-/uw/-silence (the word "two").

tional stress. Combining HNM to efficiently represent and modify speech signals with a unit selection algorithm may alleviate the difficulty of the prosodic modification problem in conventional concatenative speech synthesis systems.

In HNM, the speech spectrum is divided into two bands: a low band, which is represented by harmonically related sinusoids with slowly time-varying amplitudes and frequencies, and a high band that is instantiated by a time-varying auto-regressive (AR) model that is excited by Gaussian noise. HNM analysis consists of three steps. First, fundamental frequency (F_0) and maximum voiced frequency that determine the number of harmonics used are set using a time-domain approach. Then, harmonic amplitudes and phases are estimated by minimizing a weighted time-domain least squares criterion. Finally, the AR filter for the high band is estimated by the autocorrelation approach. The analysis windows are set at a pitch-synchronous rate during voiced portions of speech and at a fixed rate during unvoiced portions. Note that the length of local pitch epochs in HNM are estimated internally. Analysis windows are two pitch epochs long. For HNM synthesis, inter-unit phase mismatches are eliminated using the center-of-gravity approach [15]. Prosody may be altered as desired. Around unit concatenation points, the HNM parameters are smoothed in order to minimize residual discontinuities by employing a linear interpolation over a small number of frames. The actual synthesis is done by following the overlap-and-add paradigm. For each frame, the noise part is high-pass filtered according to the maximum voiced frequency found during analysis. Furthermore, the noise part is modulated by a parametric triangular envelope synchronized in time with the pitch period. Details of HNM and its application to concatenative synthesis can be found in [16].

2.1.2   FORMANT SYNTHESIS

The efficient representation of the speech signal in the spectral domain led to the development of formant (or resonance) synthesizers. In formant synthesis, the acoustic characteristics of the vocal tract are modeled directly in the frequency domain by a set of resonators. Formant synthesis is based on the assumption of source-filter separation and attempts to model the acoustics of the vocal tract using poles (formants). Formant synthesizers may also introduce additional antiresonances to accommodate zeros in the transfer function for the production of nasal and unvoiced sounds. They are also suitable for including effects of radiation at the lips and nostrils.

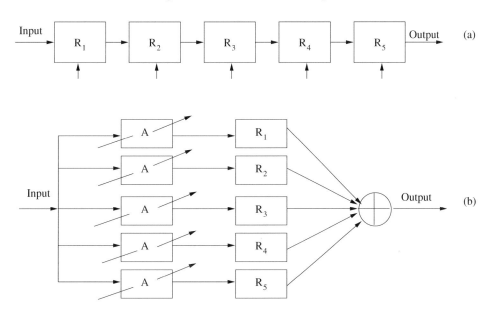

Figure 2.3: Block diagrams illustrating different structures of formant synthesizers: (a) cascade, (b) parallel.

Formant synthesizers have essentially two implementation structures. The first structure is based on the orator verbis electric (OVE) synthesizer developed by Fant [17]. It includes a set of resonators connected in cascade which models the vocal tract transfer function in the frequency domain. A simplified structure of a cascaded formant synthesizer is shown in Figure 2.3. Five resonators are selected to cover the range up to 5 kHz and are adjusted by a set of control parameters representing formant frequencies and bandwidths. The second structure of formant synthesizers is based on the notion of Lawrence and Anthony's [18], [19] parametric artificial talking (PAT) device. The resonators which simulate the vocal tract transfer function are connected in parallel as shown in Figure 2.3. This structure enables formant peaks to be adjusted individually for the production of voiced and unvoiced sounds. A refined version of this model was reported by Holmes [20] which incorporated four resonators to model the acoustics of the vocal tract and an additional one to accommodate nasalization. This was later extended by the same author to include four more filters to cover frequencies up to 8 kHz [21].

Comparison of the cascaded and parallel configurations is presented rigorously by Holmes [22]. In theory, both structures should be able to produce similar vocal tract transfer functions once the driving parameters are suitably adjusted. The main advantage of cascaded formant synthesizers is their ability to produce natural-sounding vowels without the need to control the relative amplitude of the individual formant resonators. These synthesizers, however, cannot easily accommodate changes of vocal effort. Parallel formant synthesizers, on the other hand, are capable of compensating for spectral variations due to the excitation and can produce fricatives and plosives adequately by simple control over the spectral levels in critical band regions above 3 kHz. To take advantage of both structures, Klatt introduced a hybrid synthesizer (*Klattalk*) [23] which requires 39 control parameters and employs cascaded resonators for generating voiced sounds and parallel resonators for producing fricatives. Antiresonances can be added in series to model the zeros in the spectrum [23]. The block diagram of the Klattalk formant synthesizer is shown in Figure 2.4.

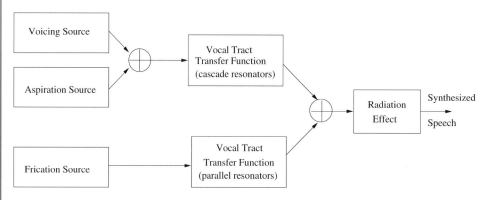

Figure 2.4: Block diagram of the Klattalk formant synthesizer.

Several models have been proposed for voiced excitation of formant synthesizers. Holmes [20] examined variations of glottal waveforms for different speakers by inverse filtering the speech signal. His finding showed that the low frequency harmonics of different excitation waveforms are similar and, perceptually, only the fundamental frequency, the *open quotient* (ratio of pulse duration to pitch period) of the pulse shape, and the *speed quotient* (ratio of rising pulse duration to falling pulse duration) are important parameters. He then recommended a glottal pulse produced by a spectral flattening procedure from the speech signal of a specific speaker. Other more sophisticated models have been reported by Fant et al. [24] and by Klatt [23]. For voiceless excitations, friction noise is introduced at the selected frequencies [23]. The ability to control the acoustic properties of the vocal tract directly by formant synthesis is important for many psychoacoustic studies. The application of formant vocoders and synthesis by rule to psychoacoustic studies has been successful.

2.1.3   ARTICULATORY SYNTHESIS

An alternative solution to generating synthetic speech is to model the physical human vocal apparatus. This is known as *articulatory synthesis*. Articulatory synthesis essentially contains two parts: a vocal fold model which represents the excitation source and a vocal tract model which describes the positions of articulators such as the tongue, lips, and jaw. It is expected that in articulatory synthesis high quality speech will be generated if the positions of the vocal apparatus are defined explicitly and an excitation source is provided which can interact appropriately with the acoustic input impedance of the vocal tract, thus providing a more exact model of the speech production mechanism. Digital synthesizers that model the speech production mechanism directly have been extensively studied in the past decades. They are essentially different in the method employed for modelling the excitation source and the vocal tract. Flanagan et al. [25] and Maeda [26] used a set of linear and nonlinear differential equations to characterize the wave propagation in the vocal system. Models which employ types of wave-digital filters have been described by Kelly and Lochbaum [27], Liljencrants [28], and Meyer et al. [29]. Sondhi and Schroeter developed a hybrid synthesizer that models the glottis in the time domain and the vocal and nasal tracts in the frequency domain [30]. Levinson and Schmidt proposed an unconstrained optimization technique to estimate the static articulatory parameters of English phonemes based on minimizing the difference between the natural speech spectra and the model speech spectra computed from the lossy Webster equation [31]. Hasegawa-Johnson and Cha have developed a low-complexity finite element model of the vocal folds, and a multispeaker MRI database of three-dimensional vocal tract shapes of vowels [32].

The advantages of articulatory synthesis are mainly the following:

1. Nonlinear interaction can be employed between the excitation source and the acoustic input impedance of the vocal tract. Conventional speech synthesizers based on linear prediction of the speech waveform assume source-tract separability. This is unrealistic for the production of consonants such as fricatives and plosives since the position and amplitude of the excitation signals are governed by the shape of the vocal tract. In our study, we even find strong source-tract interaction in the production of voiced sounds.

2. Since the control parameters of an articulatory speech synthesizer are expected to change very slowly in time to model the movements of the human articulators, they form potential candidates for very low bit rate speech coding applications. Interpolation of these parameters over several frames of speech generates reasonable vocal tract shapes, whereas interpolation of LPC coefficients can give rise to unrealistic speech sounds.

3. For TTS applications, articulatory synthesis uses rules which naturally relate to articulator movements to deal with the problems of coarticulation and unit concatenation, which are difficult problems in formant synthesis and concatenative synthesis.

4. Theoretically speaking, articulatory synthesis has the potential of producing human-like speech sounds, although the computational complexity of constructing an ac-

curate mechanical vocal folds model and the human-like vocal tract shapes and the solution of highly nonlinear differential equations governing the sound propagation inside the vocal apparatus is still too high.

5. Articulatory synthesis offers a more direct and elegant way to synthesize personalized speech sounds. For example, in order to synthesize the speech sounds of different speakers such as a tall male, an average male, and a young boy, or sounds under different stress conditions, conventional synthesizers rely on some empirical rules and certain minimum error criteria to find a set of suboptimal parameters for the waveform modification for a specific speaker. Since most of the parameters in conventional speech synthesizers are not directly related to the physical parameters involved in human speech production, the synthesized speech waveform is unsatisfactory in many cases. In the case of articulatory synthesis, we can easily change the value of several articulatory parameters such as vocal tract length and expect to synthesize personalized speech sounds very close to human sounds under different scenarios.

An articulatory synthesis system has been proposed by Rahim and Goodyear [33], [34], [35], [36] essentially based on an artificial neural network (ANN). Our work differs from their work in the following respects:

1. In Rahim's work, a multilayer perceptron (MLP) neural network was extensively used to model the nonlinear relationship between the acoustic representation of a speech signal and the vocal tract shapes. An acoustic-to-articulatory codebook was generated by the training data to cover an adequate span of the entire articulatory space of voiced and nasal sounds. Then, a dynamic programming (DP) approach was used to select an "optimal" trajectory of vocal tract shapes from the codebook by minimizing a sum of acoustical and geometrical cost components over several frames of speech. On the other hand, in our work, an advanced digital signal processing (DSP) approach is used to estimate the trajectory of articulatory parameters from the static articulatory parameters. Instead of generating a very large codebook which tries to cover the whole range of articulatory space, our approach just needs the estimation of static articulatory parameters. Then the trajectory of articulatory parameters is estimated using advanced DSP techniques. Finally, the estimated dynamic articulatory parameters are mapped into the moving vocal shapes.

2. In Rahim's articulatory synthesizer, sound propagation inside the vocal tract is assumed to be a one-dimensional plane wave. In our work, sound propagation inside the vocal apparatus is considered to be a non-plane-wave propagation inside a viscous fluid governed by the Navier-Stokes equations.

3. The glottal excitation in Rahim's synthesizer was represented by a parametric model. In our work, the excitation signal is generated by a multi-mass mechanical model of the vocal fold and a closer, and more realistic source-tract interaction is investigated.

2.2   OVERVIEW OF SPEECH PRODUCTION MODEL

2.2.1   SOURCE-FILTER SPEECH PRODUCTION MODEL

One of the most popular models in speech signal processing is the source-filter speech production model. In this model, the source waveform is filtered by a time-varying linear filter with its spectrum peaks corresponding to the vocal tract resonance. The source filter model is computationally fast and provides a vital tool for estimating basic speech parameters such as pitch, formants, and pseudo vocal tract areas. Figure 2.5 shows the system diagram of the source-filter model [37].

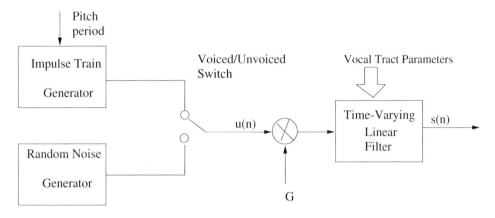

Figure 2.5: Diagram of the source-filter speech production model.

In the signal processing scenario, the source-filter model can be described by linear prediction (LP) analysis. The basic idea of LP analysis is to perform prediction of a speech sample from knowledge of past samples. The prediction error is the difference between the predicted sample and the actual one, and when minimized, leads to an optimized set of predictor coefficients. In an LP vocoder, speech is classified as either voiced or unvoiced. During voiced regions, the filter is excited by quasi-periodic pulses generated at intervals of averaged pitch periods, while for voiceless sounds, the excitation signal is provided by a white noise generator. The excitation is scaled by a gain factor G which is determined by matching the energy of original and synthetic segments of speech [37]. A complementary feature in LP analysis is that efficient and simple procedures are available for estimating the AR coefficients a_k and the gain factor G. There are two major implementations of linear prediction: the *auto-correlation* method [38] and the *covariance* method [39]. The difference between these two methods lies in their segmentation of the speech signal.

Linear prediction methods usually assume that the signal being processed is stationary within the analysis interval. It is therefore necessary to perform linear prediction over a segment of speech where the vocal tract movement is negligible. The quality of the output speech varies according to the position and size of the analysis window as well as the order of the predictive filter and the analysis method employed for estimating the prediction coefficients. A study of the variation of the prediction error with the position of the analysis window is discussed in [37]. There is usually a trade-off between the quality of the synthesized speech and the update rate of the filter parameters. For example, during regions of unvoiced speech, the filter parameters are changed at regular intervals (typically 10 ms). While for voiced segments, these parameters are updated at the beginning of each pitch period (*pitch-synchronous*) or per fixed frame size (*pitch-asynchronous*). Pitch-synchronous analysis has been found to be a more effective synthesis method, and is commonly used for estimating pseudo vocal tract area functions.

2.2.2 FRICATIVE MODEL

In his doctoral dissertation [40], Sinder proposed a fricative model which combines an existing model for acoustic propagation with a reduced model for the structure of turbulent jets. This model is based on physical principles from aeroacoustics and unsteady aerodynamics and is specifically applicable to articulatory speech synthesis.

As described in Chapter 1, fricatives are produced when air from the lungs is forced through a tight constriction in the vocal tract forming a high-speed jet. At the constriction exit, the jet becomes turbulent and generates flow-induced noise which excites resonances of the vocal tract. Sinder's fricative model is based on Howe's aeroacoustics and aerodynamics theory of sound generation [41]. In this theory, the information required to model the generation of aerodynamic sound and its propagation inside a vocal tract can be parameterized by three component models as follows:

1. *Acoustic Propagation Model:* A model which solves for acoustic wave propagation in the duct given a description of acoustic source present.
2. *Mean Potential Flow Model:* A model for the direction of the irrotational mean flow.
3. *Jet Model:* A description of the formation and convection of vorticity, including vortex strengths, speeds, and trajectories.

Figure 2.6 illustrates the components for computing aerodynamic sound generation using Howe's source term. In the following sections, we will briefly discuss the core components of Sinder's fricative model based on aeroacoustics and aerodynamics [40].

2.2.2.1 Acoustic model for unsteady potential flow

The acoustic model computes the small perturbations of velocity and pressure associated with acoustic resonance of the vocal tract. The conventional transmission line analogy is used to model plane wave propagation in the vocal tract. The transmission line T-networks include resistors representing *viscous loss*, inductors representing *fluid inertia*, and capacitors representing *fluid compliance*. A series pressure source is included in each T-network to allow the introduction of source pressure computed either from Howe's source term or a random process generator. In series, a supplemental resistance is also included to incorporate losses due to flow separation where a jet is formed. Such losses will be discussed later. Omitted are additional impedance to account for heat conduction through the wall and wall vibration. These elements generally result in slight increases in formant bandwidths,

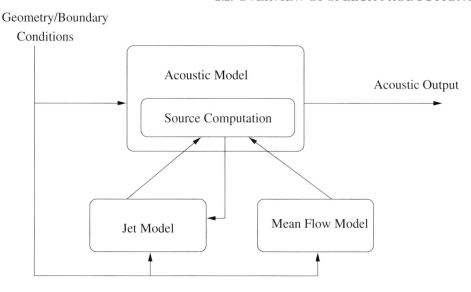

Figure 2.6: System diagram of Howe's aerodynamic sound generation model.

especially at low frequencies due to the wall vibration. Contributions to the acoustic field outside the mouth due to sound radiation through the vocal tract wall are neglected. Also, a nasal tract branch is not included since the nasal tract is not involved in fricative production. The transmission line equations are solved in the time domain in a fashion similar to the approach described in [42].

The inlet boundary condition specifies a volume source at the inlet. In speech production, lungs serve as a reservoir of air which is supplied to the vocal tract by increasing subglottal pressure. This mass source can be modulated by the valving action of the vocal folds during phonation. In the case of voiced fricatives, the inlet mass supply acts as a monopole acoustic source which excites the vocal tract. For unvoiced fricatives, a steady flow of air is supplied from the lungs, and the vocal tract is acoustically excited by a downstream acoustic source which can be modeled by a dipole acoustic source. Two inlet boundary conditions are used in Sinder's model. The first one supplies a steady volume source at a constant duct area by fixing the volume velocity there. This inlet condition is used for fixed vocal tract shapes when simulating sustained, unvoiced fricatives. The second inlet condition, which models the monopole source due to phonation, is Ishizaka and Flanagan's two-mass vocal fold model [42]. The two-mass model provides broad band excitation of the vocal tract when simulating voiced speech. When the glottis rest area is set sufficiently wide open, the folds stop vibrating and a steady volume velocity is supplied to the vocal tract. Thus, this model is used to simulate fricatives in a vowel context.

The outlet boundary condition is applied by terminating the transmission line with a radiation impedance which approximates the radiation characteristics of a piston in an infinite plane baffle [2].

Specifically, a parallel combination of a resistance R_{out} and inductance L_{out} is used in the following expressions:

$$R_{out} = \frac{128\rho c}{9\pi^2 A_{out}} \tag{2.1}$$

$$L_{out} = \frac{8\rho}{3\pi A_{out}} \sqrt{\frac{A_{out}}{\pi}} \tag{2.2}$$

where A_{out} is the area of the outlet opening.

2.2.2.2   Mean flow model for steady potential flow

The potential flow is that portion of the flow which is irrotational. In Howe's source term, the relative directions of vortex trajectories and potential flow are critical. Thus, the magnitude of the mean flow alone is not sufficient. Its direction as it flows around obstacles is a key component, which is why the mean flow model discussed below is important. For speech synthesis, a solution for arbitrary geometries is desirable, as is a solution which is computationally reduced. Therefore, rather than numerically solve a set of partial differential equations for potential flow in a fine spatial grid, a simplified model was developed. This simplified mean flow model has two major components, namely, approximation of the flow direction and the computation of streamlines.

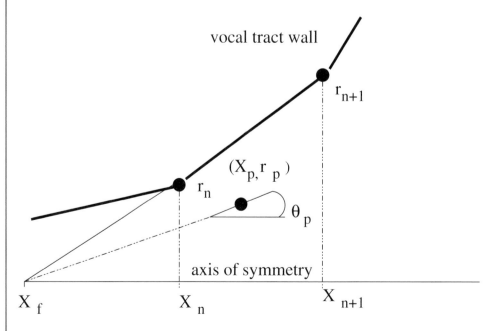

Figure 2.7: Illustration of flow direction computation for flow from left to right.

The directional information is computed by assuming that the flow uniformly expands and contracts through changes in the cross-sectional area of the vocal tract. Consider a short length of the tract as shown in Figure 2.7 [40] with endpoints x_n and x_{n+1}, radii r_n and r_{n+1}, and flow from left to right. The flow direction at an arbitrary point P at (x_p, r_p) within the small section is determined to be the same as the vector from the focus point x_f to point P. Specifically, angle θ_p between the flow direction and the axis of symmetry is given by

$$\theta_p = tan^{-1}(\frac{r_p}{x_p - x_f}) \tag{2.3}$$

$$x_f = x_n - r_n \frac{(x_{n+1} - x_n)}{(r_{n+1} - r_n)} \tag{2.4}$$

In fluid mechanics, lines which are everywhere tangent to the velocity vector are called *streamlines* [43]. The approximate streamlines can be computed by assuming that all streamlines have the same focus point within the section. Note that the distance between streamlines can be interpreted in terms of flow velocities. That is, denser streamlines indicate higher particle velocities.

2.2.2.3   Jet model for rotational flow

Fluid flow can be decomposed into irrotational and rotational components. We will briefly describe Sinder's model for the rotational component, its structure and evolution, in this section. Note that this model is a highly reduced description which still remains faithful to the physical principles important for noise generation.

At a sudden expansion in a duct, high-speed flow can separate from the relatively stagnant fluid on the downstream side of the expansion wall, resulting in a turbulent jet. Vorticity, initially present in the wall boundary layer at the jet exit, causes the jet to mix with the surrounding fluid reservoir through the action of viscous and turbulent diffusion. At first, this mixing occurs along the boundary of the jet in a shear layer which grows approximately linearly with distance from jet formation. In an axisymmetric jet, the shear layer is annular in shape. Eventually, the shear layer merges with itself, closing the annulus. At this point, the potential core of the jet ends, and the jet is fully developed. Figure 2.8 illustrates this structure [40].

The shear layer is dominated by the largest vortices (also referred to as *coherent structures*) which control the mixing rate and are nearly as wide as the shear layer itself. These large eddies shed from a location near the jet exit at nearly regular intervals, forming an array of vortex rings which convect downstream. This arrangement is referred to as a *vortex street*. The jet model is based on the structure described above, with particular focus on large eddy structures of the vortex street concept which play an important role in sound generation. This model has three major components, namely, vortex formation, vortex strength and velocity, and vortex motion and dissipation.

The separation points for vortex formation are determined through automatic shedding criteria. The criteria contain three conditions: (1) the minimum area (between 0.001 cm^2 and 0.4 cm^2), (2) the Mach number (greater than 0.01), and (3) the first two conditions exist for at least 2 ms. The

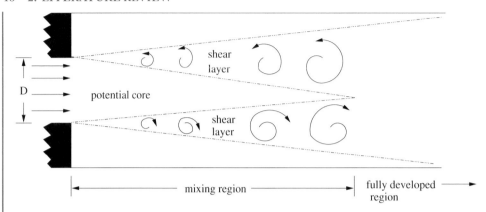

Figure 2.8: Illustration of streamline computation for flow from left to right.

time interval between shed vortices is modeled as a random variable T_{shed} with its mean value given by

$$E\{T_{shed}\} = \frac{D}{St\,U_j} \tag{2.5}$$

where D is the diameter of jet exit, St is the Strouhal number, and U_j is the jet velocity, respectively. The vortex strength and the velocity are given by

$$\Gamma = 0.5U_j^2 T_{shed} \tag{2.6}$$

$$\vec{v}_{new} = (1 - \alpha)\vec{v} + \alpha\frac{v_x}{\hat{w}_x}\hat{w} \tag{2.7}$$

where Γ denotes the vortex circulation, $\vec{v}_{new}$ denotes the velocity of the new vortex, $\vec{v}$ denotes the velocity of the previous vortex at axial location x, $\hat{w}$ denotes a unit vector with its direction aligned with the vocal tract wall at axial location x, and v_x and w_x denote the components of $\vec{v}$ and $\hat{w}$ in the axial direction, respectively. The parameter α is the cosine of the angle between the local wall and the direction of vortex velocity $\vec{v}$.

Vortices in the flow are managed in a module called vortex tracker, which associates each vortex with its own location, velocity, strength, and lifetime. The velocity and strength of each vortex is computed in Equations (2.6) - (2.7). The lifetime is specified as the axial distance the vortex travels and is experimentally determined to be $4D$. Since the vortices move much slower than the sound propagation, a first-order approximation is sufficient for updating their positions. The vortex location $\vec{x}_n$ at iteration n is given by

$$\vec{x}_n = \vec{x}_{n-1} + \vec{v}_{n-1}T \tag{2.8}$$

Finally, the aeroacoustic source contribution to the acoustic field is computed at each iteration by computing the contribution due to each vortex contained within the vortex tracker. The sound pressure p due to each individual vortex is given by

$$p = \frac{-\rho_0}{A} \int_0^{2\pi} [\vec{\Gamma} \times \vec{v} \cdot \hat{U}] d\theta = \frac{-2\pi r_\omega \rho_0}{A} [\vec{\Gamma} \times \vec{v} \cdot \hat{U}] \tag{2.9}$$

where A is the vocal tract area at the vortex location, r_ω is the radius of the vortex ring, $\hat{U}$ is a unit vector with the mean flow direction, ρ_0 is the ambient density of air, the vector $\vec{\Gamma} = \Gamma \hat{\omega}$ where Γ is the vortex strength defined in Equation (2.6), and $\hat{\omega}$ is a unit vector with its direction aligned with that of the vorticity vector $\vec{\omega}$. The symbols $\times$ and $\cdot$ denote the outer product operation and inner product operation, respectively. As vortices convect, they excite different tubelets along the vocal tract. Therefore, the source is both spatially and temporally distributed.

2.2.3   UNVOICED SPEECH SOUND PRODUCTION MODEL

In his recent paper [44], Krane extended Howe's solution of convective wave propagation and proposed a general unvoiced speech production model. The physics of sound production by vocal tract airflow is not only the primary mechanism of unvoiced sound production, but also a secondary source of sound in voicing. The sound propagation inside the vocal apparatus consists of two modes, namely, the *sound mode* (irrotational and compressible) and the *flow mode* (rotational and slightly compressible). The sound or propagation mode has been extensively studied using linear acoustics and signal processing techniques and results in a successful speech production model such as the source-filter model described in the previous section. On the other hand, the study of the flow or convective mode and its contribution to sound production has been neglected or avoided in the speech science literature for a long time. The traditional approach to analyzing air motion in the vocal tract suffers from the following misunderstandings. First, the flow is assumed to be irrotational and quasi-steady, and therefore the Bernoulli equation is sufficient to describe the relationship between pressure and particle velocity. Second, the acoustic excitation due to the flow is simply described by band-limited white noise. On the other hand, it is shown that in general, the characteristics of the aeroacoustic source are governed by the strength and spatial distribution of the jet velocity field, the convection speed and temporal spacing of vortical disturbances, and the axial extent and shape of the vocal tract. For turbulent jets, the shape of the vocal tract is the dominant factor in determining the spectral content of the source.

Krane's model provides a complete framework for issues regarding aeroacoustic source strength, spatial distribution, frequency content, and impedance. Through the study of aeroacoustics, which describes the interaction between flow mode and sound mode, Krane's model leads to three major findings. First, the key ingredient of the flow mode is the *fluid vorticity*, which acts as a means of storing air inertia as rotational motion of fluid particles. Second, due to the low Mach number and high Reynolds number of the flows in unvoiced speech production, the production of sound by speech airflow is very inefficient and in general does not affect the behavior of the flow mode.

The aeroacoustic source characteristics depend on the shape of the vocal tract and the airflow speed. Third, if the source region is acoustically compact, the source spectrum can be written as the convolution of two signals, namely, the waveform of a single vortex passing through the source region and a function describing the arrival of vortices in the source region, scaled by the strength of the particular vortex. The vortex arrival statistics determine whether the source spectrum is dominated by the vorticity field or the wall shape. If the vortex arrival time series is highly coherent periodic, it will dominate the source spectrum. On the other hand, if the vortex arrival time series is broadband, the vocal tract shape will dominate the source spectrum. More details can be found in [44].

2.3   OVERVIEW OF ARTICULATORY SPEECH MODEL

2.3.1   COKER'S MODEL

In 1976, Coker proposed an articulatory model for unlimited vocabulary speech synthesis [45]. Figure 2.9 shows the representation of the mid-sagittal section of the human vocal tract. The articulator was described by eight physical parameters: tongue body height (Y), anterior/posterior position of the tongue body (X), pharyngeal opening (P), tongue tip height (B), tongue tip curliness (R), lip opening (W), lip roundness (C), and vocal tract length (L). In 1983, Levinson and Schmidt [31] proposed an unconstrained optimization technique to estimate Coker's articulatory model parameters based on minimizing the difference between the natural speech spectra and the model spectra computed from the Webster equation. We implemented this technique on a Sun workstation to get the stationary articulatory model parameters for 52 English phone units.

2.3.2   SYNTHESIS OF SPEECH PHONEMES

In this part, we use the articulatory model to synthesize the speech signal based on the Webster equation and the overlap-and-add method. The vocal tract is modeled as soft walled acoustic tubes in which both thermal and viscous losses occur. If we assume the sinusoidal steady-state solution $p(x,t) = P(x)e^{-j\omega t}, u(x,t) = U(x)e^{-j\omega t}$, we can get the lossy Webster equation for the volume velocity in the frequency domain:

$$\frac{d^2U}{dx^2} = \frac{1}{Y(x,\omega)}\frac{dU}{dx}\frac{dY}{dx} - Y(x,\omega)Z(x,\omega)U(x,\omega) \tag{2.10}$$

where $u(x,t)$ is the volume velocity, $Z(x,\omega)$ and $Y(x,\omega)$ denote the generalized acoustic impedance and admittance per unit length, respectively. This equation can be solved by discretizing in space for different frequencies. We define $U_i^k = U(i\Delta x, k\Delta\omega)$, $Y_i^k = Y(i\Delta x, k\Delta\omega)$ where Δx and $\Delta\omega$ denote the step size of the position and frequency, respectively. After replacing the second derivatives by central differences and first derivative by backward differences, we get the following recursive equation:

$$U_{i-1}^k = \frac{1}{\frac{Y_{i-1}^k}{Y_i^k} - 2}U_{i+1}^k - U_i^k\left[3 + (\Delta x)^2 Z_i^k Y_i^k - \frac{Y_{i-1}^k}{Y_i^k}\right] \tag{2.11}$$

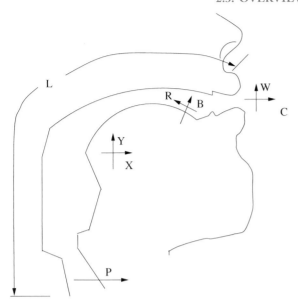

Figure 2.9: Coker's articulatory model.

where:

$$Z_i^k = \frac{jk\Delta\omega\rho}{A_i} \tag{2.12}$$

$$Y_i^k = \frac{A_i}{\rho c^2}\left[jk\omega + \frac{\omega_0^2}{\alpha + jk\Delta\omega} + \sqrt{\beta jk\Delta\omega}\right] \tag{2.13}$$

where N denotes the number of area sections, K denotes the number of frequency samples, A_i denotes the area in ith section, $\omega_0 = 406\pi$, $\alpha = 130\pi$, $\beta = 4$ and ρ denotes the media density. From Eqs. (2.12) and (2.13), we can see that $\frac{Y_{i-1}^k}{Y_i^k} = \frac{A_{i-1}}{A_i}$ and $Z_i^k Y_i^k = \frac{jk\Delta\omega}{c^2}\left[jk\Delta\omega + \frac{\omega_0^2}{\alpha + jk\Delta\omega} + \sqrt{\beta jk\Delta\omega}\right] \equiv C(k)$. We can simplify Eq. (2.11) as:

$$U_{i-1}^k = \frac{1}{\frac{A_{i-1}}{A_i} - 2}U_{i+1}^k - U_i^k\left[3 + (\Delta x)^2 C(k) - \frac{A_{i-1}}{A_i}\right] \tag{2.14}$$

Given the initial condition $U_N^k = 1$, we can compute U_0^k and thus estimate the transfer function at the frequency $k\Delta\omega$:

$$H(k\Delta\omega) = \frac{U_N^k}{U_0^k} = \frac{1}{U_0^k} \tag{2.15}$$

We can further estimate the vocal tract output by:

$$\hat{P}(k\Delta\omega) = H(k\Delta\omega)U_g(k\Delta\omega)Z_r(k\Delta\omega) \tag{2.16}$$

where $U_g(k\Delta\omega)$ and $Z_r(k\Delta\omega)$ denote the excitation and radiation load at frequency $k\Delta\omega$, respectively.

Figs. 2.10 and 2.11 show the system diagram of the articulatory synthesizer and an example of the synthesized vowel phoneme /AH/, respectively.

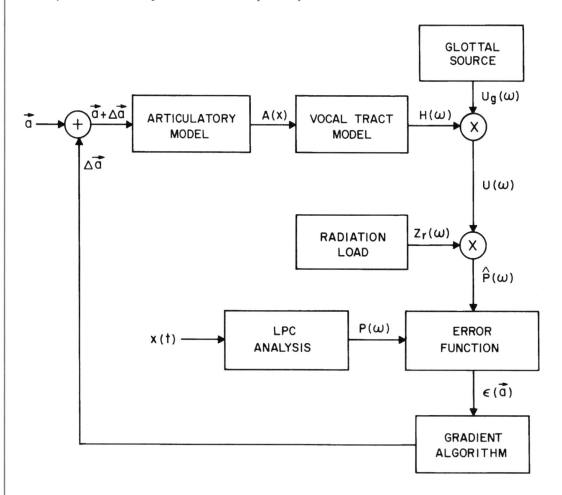

Figure 2.10: System diagram of the speech synthesizer based on lossy Webster equation and overlap-and-add method.

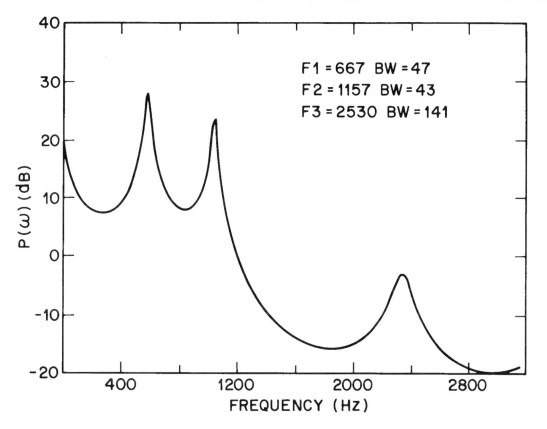

Figure 2.11: Waveform of the synthesize vowel phoneme /AH/.

2.3.3   MERMELSTEIN'S MODEL

In 1973, Mermelstein proposed an articulatory model as shown in Figure 2.12. This model permits simple control over a selected set of articulatory parameters, namely, tongue body center (C), velum (V), tongue tip (T), jaw (J), lips (L), and hyoid (H) [46]. The positions of these articulators determine the vocal tract length and cross-sectional areas, following the estimation of the vocal tract profile in the mid-sagittal plane.

Mermelstein's model consists of fixed and movable structures to enable it to articulate movements of the vocal tract during speech production. Fixed structures are approximated by straight segments and circular arcs, and represent the rear pharyngeal wall (GR), the soft palate (RM), the hard palate (MN), and the alveolar ridge (NU). These structures form the posterior-superior vocal tract outline, which is considered fixed with the exception of the region near the upper lips. Movable structures approximate the inferior-anterior vocal tract outline and are classified into two categories:

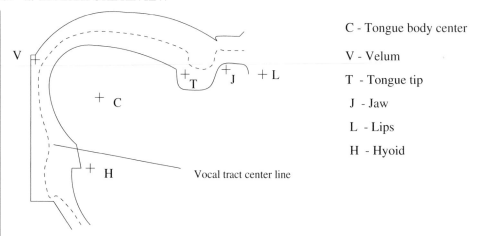

Figure 2.12: Mermelstein's articulatory model.

those whose movements are independent of other articulators (velum, hyoid bone, and jaw), and articulators whose position is a function of other articulators (tongue body, tongue tip, and lips). For example, the movement of the tongue tip is relative to that of the tongue body which itself depends on the position of the jaw. The jaw and the velum are considered movable structures with one degree of freedom. All other articulators have two degrees of freedom.

By proper selection of the various positions of the articulators, Mermelstein's model is able to generate physiologically possible shapes that span the entire articulatory space. Table 2.1 shows the limits of the articulatory parameters in Mermelstein's model [46].

No.	Name	Meaning	Lower limit	Upper limit
1	rtp	tongue radius [cm]	0.5	3.5
2	tejp	jaw angle [deg]	10.9	28.0
3	xcp	tongue center x [cm]	4.0	10.0
4	stp	tongue blade length [cm]	2.0	4.4
5	xclp	lip protrusion x [cm]	0.36	1.89
6	xhp	hyoid position x [cm]	5.20	7.60
7	ycp	tongue center y [cm]	1.48	7.48
8	telp	tongue elevation [deg]	68.2	96.8
9	cylp	lip height y [cm]	0.93	1.32
10	yhp	hyoid position y [cm]	7.43	9.83
11	vel	velum opening [cm^2]	0	2

2.3.4   TASK-DYNAMIC MODEL

During the past decades, Rubin et al. and Saltzman and Munhall in Haskins's lab have proposed a task-dynamic model for speech production [47], [48], [49]. Their model was constructed based on a mid-sagittal view of the vocal tract and a simplified kinematic description of the vocal tract's articulatory geometry. Figure 2.13 shows the mid-sagittal view of the task-dynamic model. The presented model is associated with the control of bilabial, tongue-dorsum, and lower-tooth-height constrictions. The creation and release of constrictions of differing degrees in different regions of the vocal tract are described in two steps. The first step entails defining time-invariant dynamics at the level of tract-variable coordinates. The second step is to transform the tract-variable system into an explicitly articulatory set of coordinates to describe the articulatory details required for an adequate simulation of speech production. Table 2.2 shows the relationship between tract-variables and model articulators.

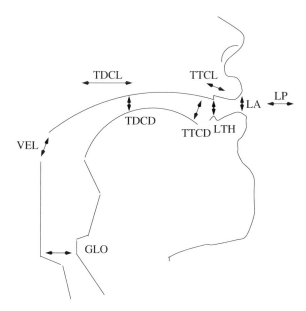

Figure 2.13: The task dynamic articulatory model, with tract variable degrees of freedom indicated by arrows.

In a more recent reference, Rubin et al. have introduce three more tract variables, namely, subglottal pressure, transglottal pressure, and delta virtual fundamental frequency to extend the previous task-dynamic model for better description of speech production. Details of the new model can be found in [50].

Table 2.2: Relationship between tract-variables and model articulators.

Name	Tract variables	Model articulators
LP	lip protrusion	upper and lower lips
LA	lip aperture	upper and lower lips, jaw
TDCL	tongue dorsum constrict location	tongue body, jaw
TDCD	tongue dorsum constrict degree	tongue body, jaw
LTH	lower tooth height	jaw
TTCL	tongue tip constrict location	tongue tip, tongue body, jaw
TTCD	tongue construct degree	tongue tip, tongue body, jaw
VEL	velic aperture	velum
CLO	glottal aperture	glottis

2.4   OVERVIEW OF THE MOTOR CONTROL OF THE ARTICULATOR

The production of speech is portrayed traditionally as a combinative process that uses a limited set of units to produce a very large number of linguistically "well-formed" utterances [51]. Segmental speech units (such as phonemes) are usually seen as discrete, static, and invariant across a variety of contexts. Putatively, such characteristics allow speech production to be generative because units of this kind can be concatenated easily in any order to form new strings. During speech production, the shape of the vocal tract changes constantly over time. These changes in shape are produced by the movements of a number of relatively independent articulators (e.g., velum, tongue, lips, jaw, etc.). Although the speech segmental units are discrete, the articulatory pattern movements of different articulators are interleaved into a continuous gestural flow.

Much theoretical and empirical evidence from the study of skilled movements of the limbs and speech articulators supports the hypothesis that significant informational units of action do not entail rigid or hard-wired control of joint and muscle variables. Rather, these units or coordinative structures must be defined abstractly or functionally in a task-specific, flexible manner. In the next section, we will discuss the motor control of the articulators according to some dynamic model or minimum cost hypothesis.

2.4.1   A DYNAMIC MODEL OF ARTICULATION

In [48], Saltzman and Munhall proposed a dynamic model of articulation with two functionally distinct but interacting levels. The *intergestural* level is defined according to both *model articulator* and *tract variable* coordinates. Tract variables (e.g., bilabial aperture) are the coordinates in which context-independent gestural "intents" are framed, and model articulators (e.g., lips and jaw) are the coordinates in which context-dependent gestural performances are expressed. Figure 2.14 shows the schematic illustration of the two-level model for speech production, with associated coordinate systems indicated. The solid arrow line from the intergestural to the interarticulator levels denotes the feedforward flow of gestural activation. The dashed arrow line indicates feedback of ongoing tract variable and model articulator state information to the intergestural level. Invariant gestural units are posited in the form of relations between particular subsets of these coordinates and sets of context-independent dynamic parameters (e.g., target position and stiffness). Contextually conditioned variability across different utterances results from the manner in which the influences of gestural units associated with the utterances are gated and blended into ongoing processes of articulatory control and coordination. The activation coordinate of each unit can be interpreted as the strength with which the associated gesture attempts to shape vocal tract movements at any given point in time. The tract variable and model articulator coordinates of each unit specify the particular vocal-tract constriction (e.g., bilabial) and set of articulators (e.g., lips and jaw) whose behaviors are directly affected by the associated unit's activation. The intergestural level accounts for patterns of relative timing and cohesion among the activation intervals of gestural units that participate in a given utterance, e.g., the activation intervals for tongue-dorsum and bilabial gestures in a vowel-bilabial-vowel sequence. The interarticulator level accounts for the coordination among articulators evident at a given point in time due to the currently active set of gestures, e.g., the coordination among lips, jaw, and tongue during periods of vocalic and bilabial gestural coproduction [48].

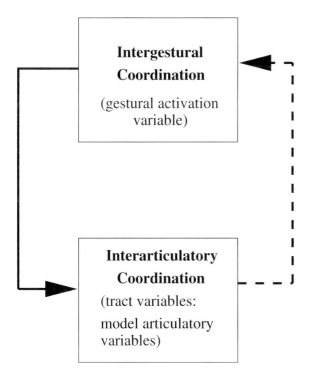

Figure 2.14: Diagram of the two-level dynamic model of speech production.

These attributes of task-specific flexibility, functional definition, and time-invariant dynamics can be incorporated into a task-dynamic model of coordinative structures. Each gesture in an utterance is associated with a corresponding tract-variable dynamic system. In this approach, all such dynamic systems are defined as tract-variable point-attractors, i.e., each is modeled by a damped, second-order linear differential equation (analogous to a damped mass-spring). The tract-variable motion equations are defined in matrix form as follows:

$$\ddot{\vec{z}} = M^{-1}(-B\dot{\vec{z}} - K\triangle\vec{z}) \tag{2.17}$$

where $\vec{z}$ is the $m \times 1$ vector of current tract-variable positions, $\dot{\vec{z}}$ and $\ddot{\vec{z}}$ are the first and second derivatives of $\vec{z}$ respect to time; M is an $m \times m$ diagonal matrix of inertial coefficients; B is an $m \times m$ diagonal matrix of tract-variable damping coefficients; K is an $m \times m$ diagonal matrix of tract-variable stiffness coefficients; and $\triangle\vec{z} = \vec{z} - \vec{z}_0$, where z_0 is the target or rest position vector for the tract variables. The components of B, K, and $\vec{z}_0$ vary during the utterance according to the ongoing set of gestures being produced. For example, different vowel gestures are distinguished in part by corresponding differences in target positions for the associated set of tongue-dorsum point attractors. Similarly, vowels and consonant gestures are distinguished in part by corresponding differences in stiffness coefficients, with vowel gestures being slower (less stiff) than consonant gestures. Thus, Equation (2.17) describes a linear system of tract-variable equations with time-varying coefficients, whose values are functions of the current active gesture set.

Furthermore, a dynamic system for controlling the model articulators is specified by expressing tract variables ($\vec{z}$, $\dot{\vec{z}}$, $\ddot{\vec{z}}$) as functions of the corresponding model articulator variables ($\vec{\phi}$, $\dot{\vec{\phi}}$, $\ddot{\vec{\phi}}$). The tract variables of Equation (2.17) are transformed into model articulator variables using the following direct kinematic relationships:

$$\vec{z} = \vec{z}(\vec{\phi}) \tag{2.18}$$

$$\dot{\vec{z}} = J(\vec{\phi})\dot{\vec{\phi}} \tag{2.19}$$

$$\ddot{\vec{z}} = J(\vec{\phi})\ddot{\vec{\phi}} + \dot{J}(\vec{\phi}, \dot{\vec{\phi}})\dot{\vec{\phi}} \tag{2.20}$$

where $\vec{\phi}$ is the $n \times 1$ vector of current articulator positions, and $J(\vec{\phi})$ is the $m \times n$ Jacobian transformation matrix whose elements J_{ij} are partial derivatives $\frac{\partial z_i}{\partial \phi_j}$ evaluated at the current $\vec{\phi}$. The elements of J and $\dot{J}$ reflect the geometrical relationships among motions of the model articulators and motions of the corresponding tract variables. Using the direct kinematic relationships in Equations (2.18) - (2.20), the equation of motion derived for the actively controlled model articulators is given by

$$\ddot{\vec{\phi}}_A = J^*(M^{-1}[-BJ\dot{\vec{\phi}} - K\triangle\vec{z}(\vec{\phi})]) - J^*\dot{J}\dot{\vec{\phi}} \tag{2.21}$$

where $\vec{\phi}_A$ is an articulatory acceleration vector representing the active driving influences on the model articulator, and $J^* = W^{-1}J^T(JW^{-1}J^T)^{-1}$ is an $n \times m$ weighted Jacobian pseudo-inverse, where W is an $n \times n$ positive definite articulatory weighting matrix whose elements are constant during a given isolated gesture. The pseudo-inverse is used because there are more model articulator variables than tract variables for this task. More specifically, using the pseudo-inverse provides a unique, optimal, least square solution for the redundant differential transformation from tract variables to model articulator variables that is weighted according to the pattern of elements in the W matrix.

2.4.2   MOTOR CONTROL BASED ON MINIMUM COST PRINCIPLES

In 1983, Nelson proposed several articulator motor control systems based on some minimum cost principles [52]. For displacement of a mass m, along the dimension x with instantaneous velocity v, the equations governing the motion are

$$\frac{dx}{dt} = v, \quad \frac{d(mv)}{dt} = f_a(t) - f_d(t) \tag{2.22}$$

where $f_a(t)$ is the net applied force along the x dimension and $f_d(t)$ is the net dissipative force (friction) resulting from the movement. We assume, for the range of displacements and velocities remaining within some linearity region L, that the mass is constant and the dissipative force is a linear function of the velocity. We further assume that there is some limit F_{max} on the magnitude of the applied force. If we now define the control action $u(t)$ as applied force per unit mass (acceleration), we can get

$$u(t) = \frac{f_a(t)}{m}, \quad |u(t)| \le U_{max} \equiv \frac{F_{max}}{m} \tag{2.23}$$

The equations of motion can be written in the normalized form

$$\dot{x}(t) = v(t), \quad x(0) = 0, \quad x(T) = D \tag{2.24}$$

$$\dot{v}(t) = u(t) - bv(t), \quad v(0) = 0, \quad v(T) = 0, \quad |u(t)| \le U_{max} \tag{2.25}$$

Because they are invariant under translation in x, Equations (2.24)–(2.25) can describe each segment of a sequence of motions along the dimension x of various distances D and movement time (durations) T between successive equilibrium states ($x = x_i$, $v = 0$). Each segment of the sequence depends only on the initial state (x_i, 0) and the control force action $u(t)$ over the current movement segment. Thus, the control strategy for each segment may be considered independently from that in the other segments.

The performance objectives can be expressed in terms of minimizing some measure of physical cost associated with accomplishing the movement. Five measures of cost that might be considered in relation to skilled movements are as follows:

$$min\{time\ cost\} = min\{T\} \tag{2.26}$$

$$min\{force\ cost = min\{A\} = min\{max_{t\in(0,T)}|u(t)|\} \tag{2.27}$$

$$min\{impulse\ cost\} = min\{I\} = min\{\frac{1}{2}\int_0^T |u(t)|dt\} \tag{2.28}$$

$$min\{energy\ cost\} = min\{E\} = min\{\frac{1}{2U_{max}}\int_0^T u^2(t)dt\} \tag{2.29}$$

$$min\{jerk\ cost\} = min\{J\} = min\{\frac{1}{2}\int_0^T \dot{a}^2(t)dt\} \tag{2.30}$$

where $\dot{a}$ is the rate of change of acceleration (jerk) of the movement. Note that the minimum-jerk cost solution will result in $x(t)$ trajectories which are equivalent to the cubic spline interpolation results. Furthermore, the $x(t)$ trajectory of the minimum-jerk cost principle is similar to the trajectories generated by the task-dynamic model described in the previous section.

2.5   SUMMARY

In this chapter, we reviewed the state-of-art techniques of TTS, speech production modelling, articulatory models, and the motor control of the articulator, which are all related to the work of this book. First, we reviewed three approaches for speech synthesis, namely concatenative synthesis, formant synthesis, and articulatory synthesis. Then we compared two speech production models, i.e., the source-filter model and the speech production model based on fluid dynamics. The source-filter model assumes source-tract separability and usually assumes one-dimensional plane wave propagation inside the human vocal tract. On the other hand, researchers such as Flanagan, Levinson, Howe, Shadle, and Krane et al. view speech production as a fluid dynamic process which includes the physical model of the human vocal apparatus and the fluid dynamic description of the wave propagation inside the vocal apparatus. In this chapter, we also described three popular articulatory models, namely, Coker's model, Mermelstein's model, and the task-dynamic model developed in Haskin's laboratory. Finally, we reviewed a mass-spring model and several minimum cost principles for the motor control of the articulator. In the next chapter, we will introduce our approach for the motor control of the articultor in our articulatory speech synthesis system.

CHAPTER 3

Estimation of Dynamic Articulatory Parameters

3.1 CUBIC SPLINE METHOD

A major difficulty in generating natural-sounding speech using articulatory synthesis is the lack of sufficient data on the motion of the articulators that control the parameters of the synthesizer. Suppose that we have N discrete phonemes with their static articulatory parameters denoted by $f_{i,j}, i = 1, 2, \cdots, N, \quad j = 1, 2, \cdots, M$, where M denotes the number of articulatory parameters per phoneme ($M = 8$ in our case). Our objective is to estimate the articulatory parameters of the K frames between consecutive phonemes i and $i + 1$. We can solve this problem by obtaining a cubic spline function which interpolates the function f at $x_0, x_1, \cdots, x_N$. In each of the subintervals $I_i = [x_i, x_{i+1}]$ of the interpolation range, s is a polynomial of degree at most three. Denoting this polynomial by s_i, we have

$$s(x) = s_i(x) \qquad x \in I_i, i = 0, 1, \cdots, N - 1 \tag{3.1}$$

A convenient formulation of s_i will be in terms of the distance of x from the starting end of the interval I_i, and so we can define the new variable u_i by

$$u_i = x - x_i \qquad i = 0, 1, \cdots, N \tag{3.2}$$

Observe that $\frac{du_i}{dx} = 1$ for every i, and so differentiation or integration with respect to x and with respect to u_i will be equivalent. We denote the step lengths between the knots by $h_i = x_{i+1} - x_i = u_i - u_{i+1}$. The conditions which must be satisfied are that s must interpolate f at $x_0, x_1, \cdots, x_N$ and $\dot{s}$, and $\ddot{s}$ must be continuous at the interior knots $x_1, x_2, \cdots, x_{N-1}$. We will begin with the continuity of $\ddot{s}$. On each of the intervals I_i, s is a cubic, and so $\ddot{s}$ is the first-degree polynomial $\ddot{s}_i$. Let us denote its values at the knots by

$$\ddot{s}(x_i) = A_i, \qquad i = 0, 1, \cdots, N \tag{3.3}$$

Since $\ddot{s}_i$ is a linear function, we have, for each i,

$$\ddot{s}_i(x) = \frac{A_{i+1}(x - x_i) - A_i(x - x_{i+1})}{h_i} = \frac{A_{i+1}u_i - A_i u_{i+1}}{h_i} \tag{3.4}$$

We may integrate Equation (3.4) twice to get

$$s_i(x) = \frac{A_{i+1}u_i^3 - A_i u_{i+1}^3}{6h_i} + cx + d \tag{3.5}$$

where c and d are constants of integration. This can be conveniently written in the form

$$s_i(x) = \frac{A_{i+1}u_i^3 - A_i u_{i+1}^3}{6h_i} - B_i u_{i+1} + C_i u_i \tag{3.6}$$

Now we should choose the coefficients of Equation (3.6) for $i = 0, 1, \cdots, N - 1$, so that both the interpolation conditions and first-derivative continuity are satisfied. Consider first the interpolation conditions. At the point x_i, we have $u_i = 0$ and $u_{i+1} = -h_i$. Denoting $f(x_i)$ by f_i and substituting these values into Equation (3.6), we get

$$f_i = \frac{A_i h_i^2}{6} + B_i h_i, \quad i = 0, 1, \cdots, N - 1 \tag{3.7}$$

$$f_{i+1} = \frac{A_{i+1}h_i^2}{6} + C_i h_i, \quad i = 0, 1, \cdots, N - 1 \tag{3.8}$$

Solving Equations (3.7) - (3.8) for B_i and C_i yields

$$B_i = \frac{f_i}{h_i} - \frac{A_i h_i}{6} \tag{3.9}$$

$$C_i = \frac{f_{i+1}}{h_i} - \frac{A_{i+1}h_i}{6} \tag{3.10}$$

The final system of equations is derived from the first-derivative continuity condition. These equations are obtained by differentiating Equation (3.6) with respect to x. We obtain

$$\dot{s}_i(x) = \frac{A_{i+1}u_i^2 - A_i u_{i+1}^2}{2h_i} - B_i + C_i \tag{3.11}$$

from which we may deduce that

$$\dot{s}_i(x_i) = C_i - B_i - \frac{A_i h_i}{2} \tag{3.12}$$

The continuity of $\dot{s}$ will be guaranteed if for every interior knot x_i, we have $\dot{s}_i(x_i) = \dot{s}_{i-1}(x_i)$ which yields the equation

$$\frac{(h_{i-1} + h_i)A_i}{2} + B_i - C_i - (B_{i-1} - C_{i-1}) = 0 \quad i = 1, 2, \cdots, N_1 \tag{3.13}$$

From Equations (3.9) - (3.10), we can get

$$B_i - C_i = \frac{(A_{i+1} - A_i)h_i}{6} - \frac{f_{i+1} - f_i}{h_i} \quad i = 0, 1, \cdots, N_1 \tag{3.14}$$

Denoting $\frac{f_{i+1} - f_i}{h_i}$ by d_i and substituting Equation (3.14) into Equation (3.13) for both i and $i - 1$, we get

$$\frac{h_{i-1}A_{i-1}}{6} + \frac{(h_{i-1} + h_i)A_i}{3} + \frac{h_i A_{i+1}}{6} = d_i - d_{i-1} \quad i = 1, 2, \cdots, N - 1 \tag{3.15}$$

Multiplying Equation (3.15) by 6 and denoting $d_i - d_{i-1}$ by $\triangle d_{i-1}$, the natural cubic spline interpolating f at $x_0, x_1, \cdots, x_N$ is obtained with the coefficients A_i satisfying the *tridiagonal* system

$$H\vec{a} = \vec{d} \tag{3.16}$$

where

$$H = \begin{bmatrix} 2(h_0 + h_1) & h_1 & 0 & \cdots & 0 \\ h_1 & 2(h_1 + h_2) & h_2 & \cdots & 0 \\ \cdots & \cdots & \cdots & \cdots & \cdots \\ 0 & \cdots & h_{N-3} & 2(h_{N-3} + h_{N-2}) & h_{N-2} \\ 0 & \cdots & 0 & h_{N-2} & 2(h_{N-2} + h_{N-1}) \end{bmatrix} \tag{3.17}$$

$$\vec{a} = [A_1 \; A_2 \; \cdots \; A_{N-1}]^T \tag{3.18}$$

$$\vec{d} = [6\triangle d_0 \; 6\triangle d_1 \; \cdots \; 6\triangle d_{N-2}]^T \tag{3.19}$$

The system described by Equations (3.16) - (3.19) is an interpolation system. The shape it assumes is the shape with minimum stored energy, and it turns out to be a natural spline.

3.2   REVIEW OF THE SIGNAL REPRESENTATION TECHNIQUES

In this section, we will review some basic ideas and definitions of the signal representation techniques.

3.2.1   INTRODUCTION

Interpolation is one of the basic operations in signal processing. It is used extensively in image reconstruction, magnetic resonance imaging, and other signal processing applications. Suppose that we are given a vector space of functions X and a set of samples from a function $x(t) \in X$; the objective of interpolation is to find an element $\hat{x} \in X$ that is the optimal approximation to $x(t)$ according to certain minimum error criteria. Traditional signal processing theory uses sinc-interpolation for band-limited functions based on Shannon's sampling theory, which is stated as follows. If the highest frequency contained in an analog signal $x_a(t)$ is $F_{max} = B$ and the signal is sampled at a rate $F_s > 2F_{max} = 2B$, then $x - a(t)$ can be exactly recovered from its sample values using the interpolation function

$$g(t) = \frac{sin(2\pi Bt)}{2\pi Bt} \tag{3.20}$$

$$x_a(t) = \sum_{n=-\infty}^{+\infty} x(n)g(t - \frac{n}{F_s}) \tag{3.21}$$

where $x(n) = x_a(\frac{n}{F_s})$ are the samples of $x_a(t)$.

In practice, most researchers use short kernel methods such as bilinear interpolation, cubic convolution, or polynomial spline interpolation, which are much more efficient to implement, especially in higher dimensions. These methods are all convolution-based in the sense that they use an interpolation model of the form

$$s_h(x) = \sum_{k \in Z} c_h(k)\varphi(\frac{x}{h} - k) \tag{3.22}$$

where h is the sampling step and $\varphi(x)$ is the basic interpolation kernel. The expansion coefficients in Equation (3.22) typically correspond to the samples of the input function $s(x)$ taken on a uniform grid: $c_h(k) = s(hk)$. Figure 3.1 shows the block diagram of the convolution-based interpolator, where the impulse response of the reconstruction filter is $\varphi_h(x) = \varphi(\frac{x}{h})$.

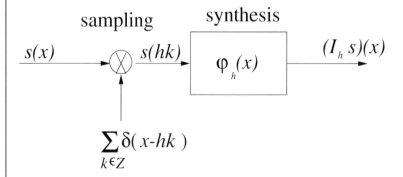

Figure 3.1: Block diagram of the convolution-based interpolator.

More recently, researchers have proposed a systematic formulation of this class of representations using the Hilbert space framework [53], [54]. The corresponding least square (LS) solution can be obtained through a simple modification of the basic interpolation procedure, which consists of applying an appropriate prefilter to $s(x)$ prior to sampling as shown in Figure 3.2. In the figure, the signal approximation $P_h s$ corresponds to the orthogonal projection of s onto the signal subspace $V_h = span\{\varphi(\frac{x}{h} - k)\}_{k \in Z}$. The impulse response of the optimal prefilter is $h^{-1} \cdot \tilde{\varphi}(-\frac{x}{h})$.

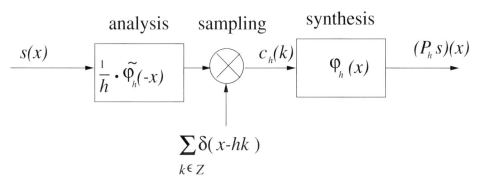

Figure 3.2: Block diagram of the convolution-based signal approximation. Block diagram of the convolution-based signal approximation.

The non-bandlimited convolution-based signal representations have a lot of advantages for signal processing such as computational efficiency and simplicity of implementation. In [53], the authors answered two fundamental questions in this signal representation. First, they provided quantitative error estimates that can be used for the appropriate selection of the sampling step h. Second, they compared three signal approximation methods – namely, least squares approximation, interpolation, and quasi-interpolation – and they derived some quantitative error bounds on these methods. Before performing the error bounds analysis of these methods, we will first introduce some notations.

3.2.2 L_2 SPACE

L_2 is the space of measurable, square-integrable, real-valued functions or signals $s(x)$, $x \in R$. It is a Hilbert space whose L_2-norm is induced from the inner product

$$< s(x), r(x) > = \int_{-\infty}^{+\infty} s(x)r(x)dx = \frac{1}{2\pi} \int_{-\infty}^{+\infty} \overline{S(\omega)}R(\omega)d\omega \qquad (3.23)$$

where $S(\omega)$ and $R(\omega)$ denote the Fourier transforms of $s(x)$ and $r(x)$, respectively. L_∞ is defined as

$$\|s\|_\infty = \lim_{p \to +\infty} [\int_{-\infty}^{+\infty} |s(x)|^p]^{\frac{1}{p}} = \sup_{x \in R} |s(x)| \qquad (3.24)$$

The class of smoothness of a signal will be specified by its distance to the Sobolev space W_2^L (or W_∞^L), which is the space of functions whose L first derivatives are defined in the L_2 (L_∞) sense.

3.2.3   CONVOLUTION-BASED SIGNAL REPRESENTATIONS

A general approach to specify continuous signal representation is to consider the class of functions generated from the integer translates of a single function $\varphi(x) \in L_2$. We can adjust the resolution by varying the sampling step h and rescaling φ accordingly. The corresponding function space $V_h(\varphi) \subset L_2$ is defined as

$$V_h(\varphi) = \{s_h(x) = \sum_{k \in Z} c_h(k)\varphi(\frac{x}{h} - k) | c_h \in l_2\} \tag{3.25}$$

where l_2 is the vector space of square-summable sequences. The only restriction on the choice of the *generation function* φ is that the set $\varphi(x/h - k)_{k \in Z}$ is a Riesz basis of $V_h(\varphi)$. This is equivalent to the condition

$$0 < A \leq a_\varphi(\omega) = \sum_{k \in Z} |\Phi(\omega + 2\pi k)|^2 \leq B < +\infty \ \ a.e. \tag{3.26}$$

where $\Phi(\omega)$ is the Fourier transform of $\varphi(x)$, and the constants A and B are the Riesz bounds. This constraint ensures that each function $s_h(x)$ in $V_h(\varphi)$ is uniquely characterized by the sequence of its coefficients $c_h(k)$.

3.2.4   INTERPOLATION AND QUASI-INTERPOLATION

The simplest way to represent a continuous signal $s(x) \in L_2$ in $V_h(\varphi)$ is to use its samples as the coefficients of the representation in Equation (3.25). The corresponding *interpolation* operator, which is schematically represented by the block diagram in Figure 3.1, is defined as

$$(I_h s)(x) = \sum_{k \in Z} s(hk)\varphi(\frac{x}{h} - k) \tag{3.27}$$

The operator I_h is bounded, provided that the input signal is sufficiently smooth; for example, $s \in W_2^1$. Note that with this definition, the samples of the signal $s(x)$ and of its interpolation $(I_h s)(x)$ are not necessarily identical. To get a true interpolation (i.e., $\forall k \in Z, s(x)|_{x=hk} = (I_h s)(x)|_{x=hk}$), we need to select a generating function $\varphi_{int} \in V_1(\varphi)$ that satisfies the interpolation property

$$\varphi_{int}(x)|_{x=k} = \delta(k) \tag{3.28}$$

where $\delta(k)$ denotes the discrete unit impulse at the origin. For a given subspace $V_h(\phi) \subset L_2$, the interpolation function $\varphi_{int}(\frac{x}{h}) \in V_h(\varphi)$ is generally unique.

Furthermore, we can relax the interpolation condition without any noticeable loss in performance. This leads to the concept of a quasi-interpolation, which is a standard notion in approximation theory, but has not yet been exploited in signal processing. By definition, a quasi-interpolant of order $L = n + 1$ is a function φ_{QI} that interpolates all polynomial $p_n(x)$ of degree n.

$$\forall p_n(x) \in P^n, \ \ \sum_{k \in Z} p_n(k)\varphi_{QI}(x - k) = p_n(x) \tag{3.29}$$

where P^n denotes the space of polynomials of degree n. The equivalent formulation of this condition in the frequency domain is

$$\Phi_{QI}(\omega)|_{\omega=2\pi k} = \delta(k) \tag{3.30}$$

$$\Phi_{QI}^{(m)}(\omega)|_{\omega=2\pi k} = 0, \quad (m = 1, \cdots, L - 1) \tag{3.31}$$

where $\Phi_{QI}(\omega)$ denotes the Fourier transform of φ_{QI} and $\Phi_{QI}^{(m)}(\omega)$ denotes the mth derivative of $\Phi_{QI}(\omega)$. In other words, $\Phi_{QI}(\xi) = 1 + O(\xi^L)$ as $\xi \to 0$. Whether or not it is possible to construct quasi-interpolants within a certain subspace $V_h(\varphi)$ depends on its order of approximation, which will be described in the Strang-Fix conditions.

3.2.5   CONVOLUTION-BASED LEAST SQUARES

A more sophisticated approach for obtaining a representation of the signal $s(x) \in L_2$ in $V_h(\varphi)$ is to determine its minimum L_2-norm approximation (orthogonal projection). This least squares approximation is given by

$$(P_h s)(x) = \sum_{k \in Z} c(k)\varphi(\frac{x}{h} - k) \tag{3.32}$$

$$c(k) = \frac{1}{h} < s(x), \tilde{\varphi}(\frac{x}{h} - k) > \tag{3.33}$$

where $\tilde{\varphi}$ is the dual of φ and is defined by

$$\tilde{\Phi}(\omega) = \frac{\Phi(\omega)}{a_\varphi(\omega)} \tag{3.34}$$

$$a_\varphi(\omega) = \sum_{n \in Z} |\Phi(\omega + 2n * \pi)|^2 \tag{3.35}$$

where $\Phi(\omega)$ and $\tilde{\Phi}(\omega)$ denote the Fourier transform of $\varphi(x)$ and $\tilde{\varphi}(x)$, respectively.

The orthogonal projection operator on $V_h(\varphi)$ can also be expressed in the more compact form

$$(P_h s)(x) = \int_{-\infty}^{+\infty} s(y)\frac{1}{h}K(\frac{x}{h}, \frac{y}{h})dy \tag{3.36}$$

where $K(x, y)$ is the reproducing kernel associated with the basic approximation space $V_1(\varphi)$:

$$K(x, y) = \sum_{k \in Z} \varphi(x - k)\tilde{\varphi}(y - k) \tag{3.37}$$

The only difference between the (quasi-)interpolation procedure and the least squares approach is the presence of the prefiltering module, which has a role similar to the anti-aliasing filter required in conventional sampling theory. In fact, if $\varphi(x) = sinc(x)$, then the optimal filter is precisely Shannon's ideal low-pass filter with the appropriate cutoff at the Nyquist frequency.

3.2.6   STRANG-FIX CONDITIONS

The Strang-Fix conditions relate the approximation power of the representation to the spectral characteristics of the generating function and are described as follows.

Let φ be a valid generating function with appropriate decay. The following statements are equivalent [55]:

(1). The function spaces $V_h(\varphi)$ reproduce polynomials of degree $n = L - 1$, which is equivalent to saying that there exists a function $\varphi_{QI} \in V_1(\varphi)$ (not necessarily unique) that is a quasi-interpolant of order L.

(2). There exists a function $\varphi_{QI} \in V_1(\varphi)$ such that

$$\forall x \in R, \quad \sum_{k \in Z} \varphi_{QI}(x - k) = 1 \tag{3.38}$$

$$\forall x \in R, \quad \sum_{k \in Z}(x - k)^m \varphi_{QI}(x - k) = 0, \quad (m = 1, \cdots, L - 1) \tag{3.39}$$

(3). $\Phi(\omega)$, which is the Fourier transform of φ, is nonvanishing at the origin and has zeros of at least multiplicity L at all nonzero frequencies that are integer multiples of 2π. Furthermore, the derivatives of $\Phi(\omega)$ are zero at the origin up to order L.

(4). There exists a constant C such that approximation error at step size h is bounded as

$$\forall s \in W_2^L, \quad inf_{s_h \in V_h(\varphi)} \|s - s_h\|_2 \leq C \cdot h^L \cdot \|s^{(L)}\|_2 \tag{3.40}$$

3.3   POINTWISE ERROR ANALYSIS

Although the Strang-Fix bound in Equation (3.40) is of considerable theoretical interest, it needs to be made more specific and quantitative to be of direct use for signal processing. In this section, we will describe the pointwise error analysis of different signal approximation methods. The proposition 1 is proved in [53]. We proved an improved error bound for interpolation in proposition 2 and an error bound for LS approximation in proposition 3. The basic tool for this pointwise analysis is the Taylor series expansion. Specifically, if the signal s is $(n + 1)$ times continuously differentiable (i.e., $s \in W_\infty^{n+1}$), we can write

$$s(y) = s(x) + (y - x)s^{(1)}(x) + \frac{(y - x)^2}{2!}s^{(2)}(x) + \cdots + \frac{(y - x)^n}{n!}s^{(n)}(x) + R_{n+1}(y) \tag{3.41}$$

$$R_{n+1}(y) = \frac{(y - x)^{n+1}}{n!} \cdot \int_0^1 (1 - \tau)^n s^{(n+1)}(\tau y + (1 - \tau)x)d\tau \tag{3.42}$$

3.3.1   INTERPOLATION ERROR

Let us consider the (quasi-)interpolation error defined by

$$s(x) - (I_h s)(x) = s(x) - \sum_{k \in Z} s(hk)\varphi(\frac{x}{h} - k) \tag{3.43}$$

Replacing $s(hk)$ by its Taylor series in Equation (3.41) with $y = hk$ and using the quasi-interpolation properties of φ, we get

$$s(x) - (I_h s)(x) = \frac{-h^L}{(L-1)!} \sum_{k \in Z} (k - \frac{x}{h})^L \varphi(\frac{x}{h} - k) \cdot [\int_0^1 (1-\tau)^{L-1} s^{(L)}(\tau hk + (1-\tau)x)d\tau] \tag{3.44}$$

The only remaining terms are those associated with the remainders of the Taylor series because φ is designed to perfectly interpolate all expansion terms up to degree $n = L - 1$. The uniform estimate of the error is given in the following proposition.

Proposition 1: If φ is a quasi-interpolant of order L with sufficient decay, then

$$\forall s \in W_\infty^L, \quad \|s - (I_h s)\|_\infty \leq C_{\varphi,L} \cdot h^L \cdot \|s^{(L)}\|_\infty \tag{3.45}$$

$$C_{\varphi,L} = \frac{1}{L!} \sup_{x \in [0,1]} \sum_{k \in Z} |x - k|^L |\varphi(x - k)| \tag{3.46}$$

Proof: See [53].

If $\varphi(x)$ decays like $O(x^{-(L+1)})$, we can improve our estimate by considering one more term in the Taylor series expansion. In this section, we will give the detailed proof of the following error bound.

Proposition 2: If φ decays like $O(x^{-(L+1)})$, then the pointwise estimate for $s \in W_\infty^{L+1}$ is given by

$$s(x) - (I_h s)(x) = -\frac{h^L}{L!} E_L(\frac{x}{h}) s^{(L)}(x) + O(h^{L+1}) \tag{3.47}$$

$$E_L(x) = (-1)^L \sum_{k \in Z} (x - k)^L \varphi(x - k) \tag{3.48}$$

Proof: Letting $y = hk$ and substituting it into Equation (3.41), we get

$$s(hk) = s(x) + \sum_{i=1}^{L-1} \frac{(hk - x)^i}{i!} s^{(i)}(x) + \frac{(hk - x)^L}{L!} s^{(L)}(x) + R_{L+1}(hk) \tag{3.49}$$

$$(I_h s)(x) = \sum_{k \in Z} s(hk) \varphi(\frac{x}{h} - k)$$

$$= \sum_{k \in Z} s(x) \varphi(\frac{x}{h} - k) + \sum_{i=1}^{L-1} \sum_{k \in Z} \frac{(hk - x)^i}{i!} s^{(i)}(x) \varphi(\frac{x}{h} - k)$$

$$+ \frac{h^L}{L!} E_L(\frac{x}{h}) s^{(L)}(x) + \sum_{k \in Z} R_{L+1}(hk) \varphi(\frac{x}{h} - k) \tag{3.50}$$

where $E_L(x)$ is defined in Equation (3.48). From the Strang-Fix condition in Equations (3.38) - (3.39), we get

$$\sum_{k \in Z} s(x) \varphi(\frac{x}{h} - k) = s(x) \tag{3.51}$$

$$\sum_{k \in Z} \frac{(hk - x)^i}{i!} s^{(i)}(x) \varphi(\frac{x}{h} - k) = 0, \quad (i = 1, \cdots, L - 1) \tag{3.52}$$

Then we can get

$$s(x) - (I_h s)(x) + \frac{h^L}{L!} E_L(\frac{x}{h}) s^{(L)}(x) = \sum_{k \in Z} R_{L+1}(hk) \varphi(\frac{x}{h} - k)$$

$$= -\frac{(h)^{L+1}}{L!} \sum_{k \in Z} (k - \frac{x}{h})^{L+1} \varphi(\frac{x}{h} - k) \cdot \int_0^1 (1 - \tau)^L s^{(L+1)}(\tau hk + (1 - \tau)x) d\tau \tag{3.53}$$

$$|s(x) - (I_h s)(x) + \frac{h^L}{L!} E_L(\frac{x}{h}) s^{(L)}(x)| \le \frac{h^{L+1}}{L!} \sum_{k \in Z} |\frac{x}{h} - k|^{L+1} |\varphi(\frac{x}{h} - k)|$$

$$\cdot \int_0^1 (1 - \tau)^L sup_{x \in R} |s^{(L+1)}(x)| d\tau \tag{3.54}$$

$$|s(x) - (I_h s)(x) + \frac{h^L}{L!} E_L(\frac{x}{h}) s^{(L)}(x)| \le h^{L+1} \|s^{(L+1)}\|_\infty \frac{1}{(L+1)!}$$

$$\cdot \sum_{k \in Z} |\frac{x}{h} - k|^{L+1} |\varphi(\frac{x}{h} - k)| = C_{\varphi, L+1} \cdot h^{L+1} \cdot \|s^{(L+1)}\|_\infty = O(h^{L+1}) \tag{3.55}$$

This leads to the pointwise estimate of s in Equation (3.47).

3.3.2   LEAST SQUARES ERROR

To simplify the analysis of error in the least squares case, we will use the reproducing kernel formalism. For the generating function φ which satisfies the decay condition $|\varphi(x)| \le K \cdot (1 + |x|)^{-M}$ $M \ge L$, an equivalent form of the conditions in Equations (3.38) - (3.39) is as follows [53]:

$$e_0(x) = \int_{-\infty}^{+\infty} K(x, y) dy = 1 \tag{3.56}$$

$$e_m(x) = \int_{-\infty}^{+\infty} (y-x)^m K(x,y)dy = 0, \quad m = 1, \cdots, L-1 \tag{3.57}$$

where the reproducing kernel $K(x,y)$ is defined in Equation (3.37). The corresponding pointwise error bound of the least squares is given in the following proposition.

Proposition 3: If φ is such that the conditions in Equations (3.56) - (3.57) are satisfied, then

$$\forall s \in W_\infty^L, \quad \|s - (P_h s)\|_\infty \leq C_{K,L} \cdot h^L \cdot \|s^{(L)}\|_\infty \tag{3.58}$$

$$C_{K,L} = \frac{1}{L!} \sup_{x \in R} [\int_{-\infty}^{+\infty} |x-y|^L |K(x,y)|dy] \tag{3.59}$$

Proof: Using Equation (3.36) and Equation (3.56), we can write the approximation error as

$$s(x) - (P_h s)(x) = \int_{-\infty}^{+\infty} [s(x) - s(y)] \frac{1}{h} K(\frac{x}{h}, \frac{y}{h})dy \tag{3.60}$$

Substituting the Taylor series Equation (3.41) into Equation (3.60) and noting from Equation (3.59) that the integers of $K(x,y)$ times difference monomials $(y-x)^m$ are zero for $m = 1, \cdots, L-1$, we get

$$s(x) - (P_h s)(x) = -\int_{-\infty}^{+\infty} R_L(y) \frac{1}{h} K(\frac{x}{h}, \frac{y}{h})dy \tag{3.61}$$

The remainder can also be written in the standard form:

$$R_L(y) = \frac{(y-x)^L}{L!} s^{(L)}(\xi) \tag{3.62}$$

where ξ is some value between x and y. This leads to the estimate

$$|s(x) - (P_h s)(x)| \leq \frac{h^L}{L!} \cdot \|s^{(L)}\|_\infty \cdot \int_{-\infty}^{+\infty} |\frac{y}{h} - \frac{x}{h}|^L \frac{1}{h} |K(\frac{x}{h}, \frac{y}{h})|dy \tag{3.63}$$

We then make the change of variable $y' = \frac{y}{h}$ and take the supremum at both sides of Equation (3.63) and get the error bound given in Equation (3.58).

The pointwise error analysis indicates that the local behavior of the error is qualitatively the same in the (quasi-)interpolation and least squares cases. Comparing Equations (3.45) and (3.58) we see that the only difference lies in the constants involved. In [53], the authors claimed that the constant $C_{K,L}$ for the least square case is smaller than the constant $C_{\varphi,L}$ for the (quasi-)interpolator case, but they did not prove this claim.

3.4 L_2 ERROR ANALYSIS

It is usually more informative to investigate the behavior of the L_2 error for quantification purposes. The most appropriate tool for this type of analysis is the Fourier transform. Before the error bound analysis, we first state a useful lemma [53].

Lemma 1: If F is M times continuously differentiable and $F^{(m)}(2\pi k) = 0$ for all $k \in Z$, $k \neq 0$ and $m = 0, \cdots, M - 1$, and if $F^{(M)}$ decays fast enough so that $\sum_k |F^{(M)}(\xi + 2\pi k)| \leq C < +\infty$, then

$$|\sum_{k \neq 0} F(\omega + 2\pi k)| \leq \frac{|\omega|^M}{M!} \sup_{|\xi| \leq \pi} |\sum_{k \neq 0} F^{(M)}(\xi + 2\pi k)|, \quad \forall |\omega| \leq \pi \tag{3.64}$$

3.4.1 L_2 ERROR OF QUASI-INTERPOLATION

We can write down Equation (3.27) in the frequency domain:

$$(\hat{I_h s})(\omega) = \sum_{k \in Z} S(\omega + \frac{2\pi}{h}k)\Phi(h\omega) \tag{3.65}$$

where $S(\omega)$, $\Phi(\omega)$, and $(\hat{I_h s})(\omega)$ denote the Fourier transforms of $s(x)$, $\varphi(x)$, and $(I_h s)(x)$, respectively. This leads to the following error decomposition in the frequency domain:

$$S(\omega) - (\hat{I_h s})(\omega) = [S(\omega)(1 - \Phi(h\omega))] - [\sum_{k \neq 0} S(\omega + \frac{2\pi}{h}k)\Phi(h\omega)] = E_1(\omega) + E_2(\omega) \tag{3.66}$$

where $E_1(\omega)$ and $E_2(\omega)$ denote the in-band and out-of-band error components, respectively. The L_2 error bound of quasi-interpolation is given by the following proposition.

Proposition 4: If φ is a quasi-interpolant of order L with sufficient decay, then

$$\forall s \in W_2^L, \|s - I_h s\|_2 \leq C_{\varphi,L} \cdot h^L \cdot \|s^{(L)}\|_2 \tag{3.67}$$

where $C_{\varphi,L}$ is defined in Equation (3.46).

The key idea of the proof of *Proposition 4* is to use the error decomposition in Equation (3.66), the Taylor series expansion of $1 - \Phi(h\omega)$, and the Schwartz inequality. The detailed proof was described in [53].

3.4.2 L_2 ERROR OF THE LS APPROXIMATION

The Fourier representation of the LS approximation of Equation (3.32) is given by

$$
\begin{aligned}
(\hat{P_h s})(\omega) &= \Phi(h\omega) \sum_{k \in Z} \overline{\Phi(h\omega + 2\pi k)} S(\omega + \frac{2\pi}{h}k) \\
&= \Phi(h\omega) \sum_{k \in Z} \frac{\overline{\Phi(h\omega + 2\pi k)}}{a_\varphi(h\omega)} \cdot S(\omega + \frac{2\pi}{h}k)
\end{aligned} \tag{3.68}
$$

where $a_\varphi(\omega)$ is defined in Equation (3.35). The approximation error in the Fourier domain can be decomposed into two components:

$$S(\omega) - (\hat{P_h s})(\omega) = e_1(\omega) + e_2(\omega) \tag{3.69}$$

$$e_1(\omega) = [1 - \frac{|\Phi(h\omega)|^2}{a_\varphi(h\omega)}]S(\omega) \tag{3.70}$$

$$e_2(\omega) = \sum_{k \neq 0} \frac{\Phi(h\omega)}{a_\varphi(h\omega)} \cdot \overline{\Phi(h\omega + 2\pi k)} \cdot S(\omega + \frac{2\pi}{h}k) \tag{3.71}$$

where $e_1(\omega)$ and $e_2(\omega)$ denote the in-band error and out-of-band error, respectively. The L_2 error bound of the LS approximation is given by the following proposition.

Proposition 5: If $\Phi^{(m)}(2\pi k) = 0$, $k \in Z$, $k \neq 0$, and $m = 0, \cdots, L-1$, then

$$\forall s \in W_2^L, \|s - P_h s\|_2 \leq K_{\varphi,2L} \cdot h^{2L} \cdot \|s^{(2L)}\|_2 + K_{\varphi,2L}^{1/2} \cdot h^L \cdot \|s^{(L)}\|_2 \tag{3.72}$$

$$K_{\varphi,2L} = \frac{1}{(2L)!} \cdot \frac{1}{A} \cdot \sup_\xi |\sum_{k \neq 0} (|\Phi|^2)^{(2L)}(\xi + 2\pi k)| \tag{3.73}$$

where $A = inf_\omega[a_\varphi(\omega)]$ is the lower Riesz bound.

The key idea of the proof of *Proposition 5* is to use the error decomposition in Equation (3.69), the Taylor expansion in *Lemma 1*, and the Schwartz inequality. The detailed proof of this proposition is given in [53].

3.4.3   COMPARISON

The L_2 bounds in Equations (3.67) and (3.72) are both consistent with the Strang-Fix conditions in Equation (3.40). However, the error bound for the LS case provides a finer characterization of the error. It consists of two distinct terms that represent the in-band (e_1) and out-of-band (e_2) contribution to the error, respectively. For a small value of h, the first part of the error becomes negligible and the bound is dominated by the second $O(h^L)$ term. The corresponding constant $K_{\varphi,2L}^{1/2}$ turns out to be smaller than the constant C in the Strang-Fix bound in Equation (3.40)., or $C_{\varphi,L}$ in Equation (3.67). This is a first indication that there is an advantage in using least squares over interpolation. In addition, we note that in the case of larger values of h, the first term in the LS error bound becomes dominant and it has the characteristic form of the error of an interpolator of order $2L$. We further observe that the function $\Phi_{2L}(\omega) = \frac{|\Phi(\omega)|^2}{a_\varphi(\omega)}$ represents the frequency response of an interpolator of order $2L$. Therefore, under the condition that $\|e_1\| \gg \|e_2\|$, the LS solution of order L should perform as well as the corresponding interpolator with twice the order. This condition typically arises for larger h when the signal is somewhat undersampled. In the following section, we will apply this observation to a specific task where the observed discrete signal is undersampled.

3.5   EXPERIMENTAL RESULTS

Suppose we are given N discrete phonemes with their static articulatory parameters denoted by $s_m(n)$, $m = 1, 2, \cdots M$, $n = 1, 2, \cdots, N$, where M denotes the number of articulatory parameters per phoneme ($M = 8$ in our case). Our objective is to estimate the dynamic articulatory parameters between consecutive phonemes. This problem can be restated as given N discrete observations $s_m(n)$ of M continuous functions $s_m(x)$, we want to construct the functions $\hat{s}_m(x)$ which is an "optimal" approximation to $s_m(x)$ according to certain criteria. Due to the physical constraints of the human vocal apparatus, all of its parameters, such as tongue body height, have finite values. Further, the articulatory trajectories are continuous over time, and their L^{th} derivatives also have finite value ($L \geq 2$). Thus, the functions $s_m(x) \in W_2^L (W_\infty^L)$, and we can use the interpolation or LS approximation discussed above to solve the dynamic parameter estimation problem. In this section, we investigate the cubic spline interpolation and the LS approximation, which is realized by an interpolation followed by an "optimal" filtering.

There are many kernel functions to choose for a practical LS approximation. In our system, we design the kernel function according to the following criterion. Given N discrete observations $s_m(n)$ of M continuous functions $s_m(x)$, we want to construct the functions $\hat{s}_m(x)$ which will minimize the combination of the mean square errors at the target points and the acceleration of the trajectory function defined by

$$min \ \{\sum_{n=1}^{N} p_n(s_m(n) - \hat{s}_m(n))^2 + \int_{x_1}^{x_N} [\ddot{\hat{s}}_m(x)]^2 dx\} \tag{3.74}$$

The first term in the right-hand side of Equation (3.74) denotes the weighted sum of the square error at the target points. The second term in the right side of Equation (3.74) denotes second order differentiation of the approximation function $s_m(x)$. The physical meaning of this criterion is to minimize both the weighted error at target points and allow minimal abrupt changes at the articulatory trajectory. We call this the combined minimum error and minimum jerk criterion.

The minimization of Equation (3.74) under the continuity constraints of $\dot{\hat{s}}_m(x)$ and $\ddot{\hat{s}}_m(x)$ gives

$$A\vec{d} = C\hat{\vec{s}}_m \tag{3.75}$$

$$\hat{\vec{s}}_m = \vec{s}_m - P^{-1}C^T\vec{d} \tag{3.76}$$

where $\vec{s}_m = [s_m(1) \quad s_m(2) \quad \cdots \quad s_m(N)]^T$, $\hat{\vec{s}}_m = [\hat{s}_m(1) \quad \hat{s}_m(2) \quad \cdots \quad \hat{s}_m(N)]^T$, $\vec{d} = [\ddot{\hat{s}}_m(x_2) \quad \ddot{\hat{s}}_m(x_3) \quad \cdots \ddot{\hat{s}}_m(x_{N-1})]^T$, $P = diag[p_1 \quad p_2 \quad \cdots p_N]$. Matrix A is an $(N-2) \times (N-2)$ symmetric three-diagonal matrix with $\frac{h_i}{6}$, $\frac{(h_i + h_{i+1})}{3}$, and $\frac{h_{i+1}}{6}$ as the nonzero elements of the ith row or column. Matrix C is an $(N-2) \times N$ three-diagonal matrix with $\frac{1}{h_i}$, $(\frac{-1}{h_i} - \frac{-1}{h_{i+1}})$, and $\frac{1}{h_{i+1}}$ as the nonzero elements of the ith row. If we assume equal-spaced data with equal weights, i.e.,

$p_i = \frac{1}{p}$, then the matrices A and C are simplified as

$$A = \frac{1}{6} \begin{bmatrix} 4 & 1 & & & 0 \\ 1 & 4 & \ddots & & \\ & \ddots & \ddots & 1 \\ 0 & & 1 & 4 \end{bmatrix} \tag{3.77}$$

$$C = \begin{bmatrix} 1 & -2 & 1 & & & 0 \\ & 1 & -2 & 1 & & \\ & & \ddots & \ddots & \ddots & \\ 0 & & & 1 & -2 & 1 \end{bmatrix} \tag{3.78}$$

From Equations (3.75) and (3.76) we can get

$$(A + B)\hat{\vec{s}}_m = \vec{s}_m \tag{3.79}$$

$$B = \frac{1}{p}CC^T \tag{3.80}$$

From Equations (3.77) and (3.78), we know $A + B$ is a symmetric five-diagonal Toeplitz matrix. Equation (3.79) means that except for the first two rows and the last two rows of $A + B$, there exists a linear relation between $\hat{\vec{s}}_m$ and $\vec{s}_m$ which can be expressed as

$$[A(z) + B(z)]\hat{\vec{s}}_m = C(z)\vec{s}_m \tag{3.81}$$

where

$$A(z) = \frac{1}{6}(z + 4 + z^{-1}) \tag{3.82}$$

$$B(z) = \frac{1}{p}(z^2 - 4z + 6 - 4z^{-1} + z^{-2}) \tag{3.83}$$

$$C(z) = z - 2 + z^{-1} \tag{3.84}$$

As a result, the Z-transform of the resulting optimal filter is given by

$$H(z) = \frac{A(z)}{A(z) + B(z)} = \frac{\frac{1}{6}(z + 4 + z^{-1})}{\frac{1}{p}\left[z^2 + (\frac{p}{6} - 4)z + (\frac{2p}{3} + 6) + (\frac{p}{6} - 4)z^{-1} + z^{-2}\right]} \tag{3.85}$$

In our experiment, the static articulatory parameters for each English phoneme were estimated using the acoustic-to-articulatory inversion algorithm proposed by Levinson and Schmidt [31]. The speech signals were recorded through a high-quality microphone in a sound booth. The analog signals were low-pass filtered at 5 kHz and sampled to 12 bits at a 10-kHz rate. The phonemes were excised from words spoken in isolation by four speakers. We then used the dynamic articulatory parameter estimation algorithm discussed in this chapter to estimate the trajectory of articulatory parameters for dipthongs and continuous English sentences.

Figures 3.3–3.10 show the estimated dynamic articulatory parameters (X: anterior/posterior position of the tongue body, Y: tongue body height, R: tongue tip curliness, B: tongue tip height) of the English sentences *Where were you while we were away?* and *Guess the question from the answer*, respectively. For each figure, the circle points denote the static articulatory parameters (phoneme targets) computed from the method in [31]. The phonetic transcription of the 17 phoneme targets of sentence *Where were you while we were away?* are /W/ /EH/ /R/ /W/ /UH/ /Y/ /OO/ /W/ /AI/ /L/ /W/ /EE/ /W/ /UH/ /A/ /W/ /AY/. The phonetic transcription of the 21 phoneme targets of sentence *Guess the question from the answer* are /G/ /EH/ /S/ /DH/ /A/ /K/ /W/ /EH/ /SH/ /I/ /N/ /F/ /R/ /AH/ /M/ /DH/ /EE/ /AA/ /N/ /S/ /ER/. The dashed line denotes the approximated continuous articulatory parameters using the cubic spline interpolation method; and the solid line denotes the estimated continuous articulatory parameters using the LS approximation approach. We can see from these figures the LS approach achieves smoother approximation of the articulatory parameter trajectory over time $(s_m(t))$ and does not have the "overshoot" problems of the cubic spline methods. One hypothesis in human speech production assumes that articulation is planned in order to minimize the abrupt movements of neuro-muscular actions. We can see from these figures that the LS solution tends to be consistent with this hypothesis and the minimum target points to error criterion because its trajectory doesn't have the abrupt changes which exist in some regions of the cubic spline solutions and it will also achieve an almost exact match at the target points on the articulatory trajectory.

We can see from this figure that the LS approximation method can successfully estimate the articulatory trajectory and thus the dynamic vocal tract movements. Although the vocal tract shape of the phoneme /EH/ is unchanged with time when produced in isolation, we can see from this figure that the vocal tract geometry of /EH/ will change dynamically with respect to its neighbor phonemes and thus result in the change of its acoustic properties at different time frames, which is called *coarticulation* in phonetics. Once we can get a good estimate of articulatory parameter trajectories, and thus of the dynamic vocal tract movements, we can apply these dynamic geometries as the boundary conditions for solutions of the Navier-Stokes equations and then synthesize highly intelligible, natural sounding continuous speech.

3.6   DISCUSSION

Signal approximation in the framework of the Hilbert space [53] provides a powerful tool in signal processing because it is unbandlimited, more efficient to implement in high dimensions, and has more quantitative error bounds than the conventional interpolation technique based on Shannon's sampling theory. The pointwise error analysis indicates that the (quasi-)interpolation and LS ap-

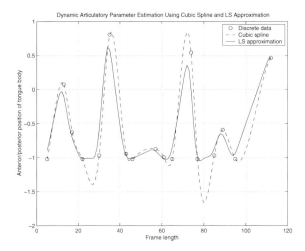

Figure 3.3: Estimated articulatory parameter X (anterior/posterior position of the tongue body) using cubic spline interpolation and LS approximation of the sentence *Where were you while we were away?*

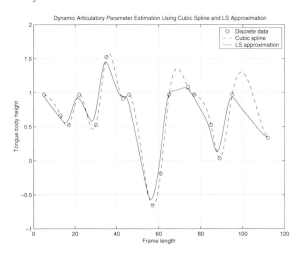

Figure 3.4: Estimated articulatory parameter Y (tongue body height) using cubic spline interpolation and LS approximation of the sentence *Where were you while we were away?*

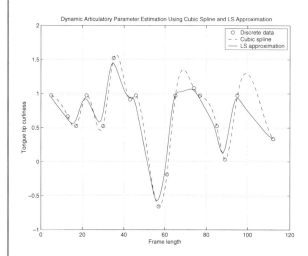

Figure 3.5: Estimated articulatory parameter R (tongue tip curliness) using cubic spline interpolation and LS approximation of the sentence *Where   were   you   while   we   were   away?*

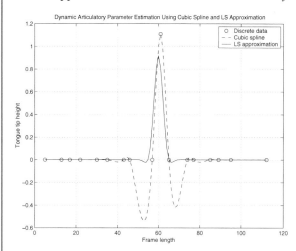

Figure 3.6: Estimated articulatory parameter B (tongue tip height) using cubic spline interpolation and LS approximation of the sentence *Where   were   you   while   we   were   away?*

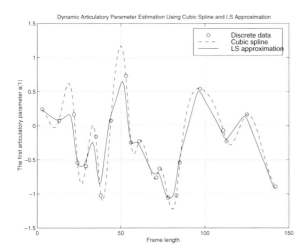

Figure 3.7: Estimated articulatory parameter X (anterior/posterior position of the tongue body) using cubic spline interpolation and LS approximation of the sentence *Guess the question from the answer.*

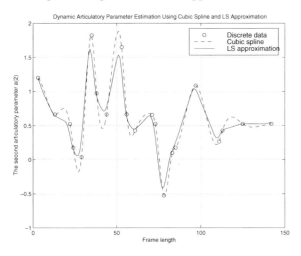

Figure 3.8: Estimated articulatory parameter Y (tongue body height) using cubic spline interpolation and LS approximation of the sentence *Guess the question from the answer.*

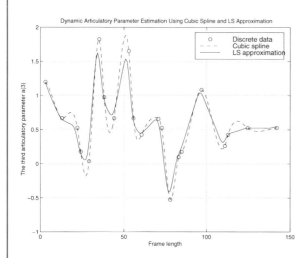

Figure 3.9: Estimated articulatory parameter R (tongue tip curliness) using cubic spline interpolation and LS approximation of the sentence *Guess   the   question   from   the   answer.*

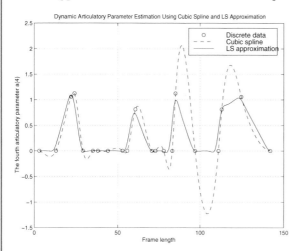

Figure 3.10: Estimated articulatory parameter B (tongue tip height) using cubic spline interpolation and LS approximation of the sentence *Guess   the   question   from   the   answer.*

proximation are qualitatively the same. The only difference is the constant terms, which tend to be smaller for the LS case. For small values of the sampling step h, the in-band component of the L_2 error of the LS approach becomes negligible and the bound is dominated by the out-of-band error term ($O(h^L)$), and the LS error still tends to be smaller than the interpolation error in this case. For larger values of h (undersampling), the first term in the LS error bounds becomes dominant and it has the characteristic form of the error of an interpolator of order $2L$. When the sampling step h approaches zero, both error bounds approach zero at the same rate. Both the spline interpolation and LS approximation approach can be successfully applied to estimate the continuous articulatory parameter trajectory given the static parameters of the discrete phonemes. Experimental results indicate that the LS approach achieves smoother estimation and does not have the "over-shooting" problem of the spline interpolation methods.

From the discussion with Prof. Hasegawa-Johnson, we know that many theories of speech production hypothesize that phoneme targets are the most extreme positions of the articulators, and that during actual speech production, the articulators spend most of their time in between phoneme targets. In other words, the articulatory trajectories should only undershoot a phoneme target. The phoneme overshoot that one observes in spline-interpolated trajectories is exactly opposite to the above hypothesis. The combined minimum error and minimum jerk criterion proposed in this book overcomes the over-shooting problem with spline interpolation.

3.7   FUTURE WORK

Traditional articulatory models use 7–15 physical parameters to model the mid-sagittal section of the human vocal tract and use certain error minimization criteria to estimate the articulatory parameters from the measured acoustical parameters (usually called the inverse problem) [45], [46], [48], [56], [57], [58]. The disadvantage of these types of articulatory models is that they are basically 2-D models and it is difficult to overcome the ambiguity problem because there is a nondenumerable infinity of area functions corresponding to a given set of acoustical parameters (such as formant frequencies). Furthermore, these articulatory models can not get fine details of the sharp constriction geometry of the 3-D vocal tract, which are important for the analysis of airflow inside the vocal apparatus. An alternative is to directly measure the vocal tract shape using X-ray or magnetic resonance imaging (MRI) techniques. The disadvantage of this approach is that it will produce a very high-dimensional signal space. For example, a typical MRI image of an English phoneme contains thousands of data points, making it impossible for direct analysis. However, a number of studies suggest that articulatory movements in speech can be approximately modeled using a few elementary articulatory parameters [59], [60], [61]. One approach is to use the latent variable model-based method to reduce the dimensionality of the MRI vocal tract shape data. The objective is to construct an articulatory speech model with a few parameters which still can represent the fine details of the 3-D vocal tract shape for different phonemes.

The assumption of a latent variable model is that the observed high-dimensional data are generated from an underlying low-dimensional process. The high dimensionality arises for several reasons, including stochastic variation and the measurement process [62]. The objective is to learn the low-dimensional generating process (defined by a small number of *latent variables* or *hidden causes*) along with a noise model, rather than directly learning a dimensionality reduction mapping. A latent variable model consists of the following three parts:

     1. the functional form of the prior in latent space,

     2. the smooth mapping from latent space to data space, and

     3. the noise model in data space.

3.8   SUMMARY

In this chapter, we presented our approach for the motor control of the articulator during continuous speech production. First, we introduced some fundamentals of the signal approximation theory. Then we stated the point error analysis and L_2 error analysis of interpolation and LS approximation. Then we proposed a combined minimum error and minimum jerk criterion for the estimation of articulatory trajectories. We also compared our results with the cubic spline interpolation approach. Simulation results showed that our approach overcomes the over-shooting problem with spline interpolation. Finally, some future work in this area was suggested.

CHAPTER 4

Construction of Articulatory Model Based on MRI Data

4.1 PROBLEM FORMULATION

A qualitative physical picture of speech sound production has developed over a long time. Fant and Flanagan [17], [2] first noted the necessity of turbulent airflow for producing unvoiced and voiced sounds. Stevens [60] incorporated many ideas from the aeroacoustics literature and concluded that the sound is produced not where the turbulent flow is formed, but instead where that turbulent flow interacts with an obstacle such as teeth in the vocal apparatus. Shadle strongly suggested that the sound produced by airflow in the vocal tract is sensitive to the three-dimensional details of vocal tract geometry, so that a simple axial area distribution may not be enough to characterize the vocal tract for the purpose of producing speech sounds.

Traditional articulatory models use 7–15 physical parameters to model the mid-sagittal section of the human vocal tract and use certain error minimization criterion to estimate the articulatory parameters from the measured acoustical parameters (usually called the inverse problem) [45], [56], [46], [57], [48], [58]. The disadvantage of these types of articulatory models is that they are basically 2-D models and it is difficult to overcome the ambiguity problem because there is a non-denumerable infinity of area functions corresponding to a given set of acoustical parameters (such as formant frequencies). Furthermore, these articulatory models can not get fine details of the sharp constriction geometry of the 3-D vocal tract which is important for the analysis of airflow inside the vocal apparatus. An alternative is to directly measure the vocal tract shape using X-ray or magnetic resonance imaging (MRI) techniques. The disadvantage of this approach is that it will produce a very high dimensional signal space. For example, a typical MRI image of an English phoneme contains thousands of data points and thus make it impossible for direct analysis. However, a number of studies suggest that articulatory movements in speech can be approximately modeled using a few elementary articulatory parameters [59], [60], [61]. In this chapter, we propose a latent variable model-based method to reduce the dimensionality of the MRI vocal tract shape data. Our objective is to construct an articulatory speech model with a few parameters which still can represent the fine details of the 3-D vocal tract shape for different phonemes.

4.2   VOCAL CORDS MODELS

One of the important components of all speech synthesis systems is the voiced sound excitation source, which models the behavior of the vocal cords. A model for voiced sound excitation of the vocal system is shown in Figure 4.1. On the left side of the figure, the air reservoir represents the lungs. An air flow with subglottal pressure P_s is expelled through the glottal orifice with volume velocity U_g. This air flow produces a local Bernoulli pressure. The subglottal pressure and the Bernoulli pressure cause the virtation of vocal cords. Here the vocal cords are represented as a mechanical oscillator, composed of mass M, spring K and viscous damping, B. The vocal tract and the nasal tract are modeled as tubes whose cross sectional areas change with distance. The acoustic volume velocities at the mouth and the nostrils are U_m and U_n respectively. The sound pressure in front of the mouth is approximately estimated by the linear superposition of the time derivatives $\dot{U}_m$ and $\dot{U}_n$.

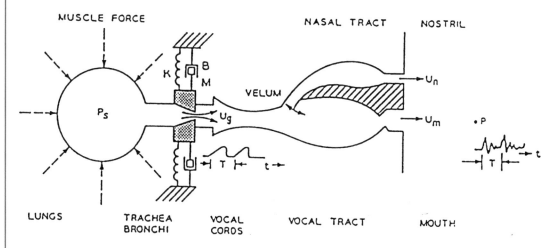

Figure 4.1: Human vocal apparatus.

The first mechanical vocal cords model was proposed by Flanagan [2], where the vocal cords are represented by a single mass. However, the one-mass model is too simple to produce some physiological details in the vocal cords. A two-mass vocal cords model was proposed later by Ishizaka and Flanagan [42]. It consists of two damped self-oscillating masses connected by a coupling spring. To improve the quality and the naturalness of the synthesized speech, a modified two-mass vocal cords model was proposed by Pelorson et al [85]. Instead of using rectangular model, two point masses drive a two-parameter curve describing the smoothness of the vocal fold surface. This smooth model allows description of unsteady flow separation with the glottis, replacing the assumption that the air flow through the glottal channel separates from the walls at a fixed point.

4.3   MULTI-MASS MODEL

To better understand the behavior of the air flow through the vocal cords, full numerical solutions are very important. The Navier-Stokes equations provide a means of solving for the behavior of the air flow in 3-Dimensions. Without considering the behavior of the air flow at each point in the vocal cords, the one-mass and two-mass models are insufficient to combine with the Navier-Stokes equations to provide the full numerical solutions. There, a multi-mass model is proposed.

The multi-mass model is shown in Figure 4.2. Similar to the one-mass and two-mass models, it is composed of self-oscillating masses. The vocal cords are divided into N equal parts. Each part is considered as a self-oscillator, with mass m_i, stiffness k_i, and damping R_i. The adjacent masses are connected by coupling springs with stiffness k_{ij}. The multi-mass model is bilaterally symmetric; therefore, only one side of vocal cords needs to be considered.

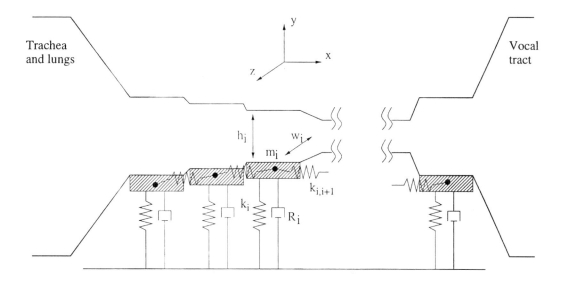

Figure 4.2: Multi-mass vocal fold model.

The system can be described by N differential equations.

$$
\begin{aligned}
F_1 &= m_1\ddot{y}_1 + R_1\dot{y}_1 + k_1 y_1 + k_{12}(y_1?y_2) \\
F_2 &= M_2\ddot{y}_2 + R_2\dot{y}_2 + k_2 y_2 + k_{23}(y_2?y_3) + k_{12}(y_2?y_1) \\
&\vdots
\end{aligned}
$$

$$F - i \quad = \quad m_i \ddot{y}_i + R_i \dot{y}_i + k_i y_i + k_{ii+1}(y_i ? y_{i-1}) + k_{i-1,i}(y_i ? u_{i-1})$$

$$\vdots$$

$$F_N \quad = \quad m_N \ddot{y}_N + R_N \dot{y}_N + k_N y_N + k_{N-1,N}(y_N ? y_{N-1})$$

$$(4.1)$$

Each force F_i is a function of local acoustic pressure p_i, which in turn is a function of subglottal

volume sped U_g. The relationship between p_i and U_g can be described by the Bernoulli equation

$$P_i \quad = \quad p_{Bernoulli} + p_{viscous} \tag{4.2}$$

$$P_i \quad = \quad \frac{\rho U_g^2}{2h_i^2 w_i^2} + \frac{12\mu \Delta x U_g}{w_i h_i^3} \tag{4.3}$$

Where ρ is the air density, μ is the dynamic viscosity coefficient, h_i is the height of the glottis, w_i is the width of the glottis, and Δx is the small increment in the x direction (see Figure 4.2). A more accurate relationship between p_i and U_g can be described by the Navier-Stokes equations.

These N 2nd order differential equations can be solved using an ordinary differential equation (ODE) solver by setting

$$v_i \quad = \quad y_i$$
$$w_i \quad = \quad \dot{v}_i = \dot{y}_i$$

With these substitutions the N 2nd order differential equations degrade to $2N$ 1st order differential equations as follows:

$$F_1 \quad = \quad m_1 \dot{w}_1 + R_1 w_1 + k_1 v_1 + k_{12}(v_1 v_2)$$
$$\dot{v}_1 \quad = \quad w_1$$
$$F_2 \quad = \quad m_2 \dot{w}_2 R_2 w_2 + k_2 v_2 + k_{23}(v_2 v_3) + k_{12}(v_2 v_1)$$
$$\dot{v}_2 \quad = \quad w_2$$
$$\vdots$$
$$F_i \quad = \quad m_i \dot{w}_i + R_i w_i + k_i v_i + k_{i,i+1}(v_i v_{i+1}) + k_{i-1,i}(v_i u_{i-1})$$
$$\dot{v}_i \quad = \quad w_i$$
$$\vdots$$
$$F_N \quad = \quad m_N \dot{w}_N + R_N w_N + k_N v_N + k_{N-1,N}(v_N v_{N-1})$$
$$\dot{v}_N \quad = \quad w_N.$$

The parameters of this model, k_i, R_i, k_{ij}, can be nonlinear. When N is large, those parameters can be approximately linear, or piecewise linear (only need to consider the open and closed condition).

4.4   SIMULATION RESULT AND FUTURE WORK

A simple simulation was performed on a ten-mass model. Assuming the force is applied only on the first mass, therefore, the force is equal to the subglottal pressure. In the simulation, the subglottal pressure is exponentially increased from zero to its maximum value.

The model parameters are linearly decreased from the first mass to the last mass. The open condition and the closed condition are treated differently. The parameters of the first and the last mass are shown in Table 1.

	open condition	*closed condition*
m_1	0.17g	0.17g
m_{10}	0.03g	0.03g
k_1	$80 N M^{-1}$	$320\ N M^{-1}$
k_{10}	$8 N M^{-1}$	$32 N M^{-1}$
R_1	$2.33 \times 10^3 N s m^{-1}$	$2.57 \times 10^{-3} N s m^{-1}$
R_{10}	$1.86 \times 10^{-3} N s m^{-1}$	$4.96 \times 10^{-3} N s m^{-1}$

The system is solved by Matlab Ordinary Differential Equation (ODE) solver. The resulting volume velocity is plotted in Figure 4.3.

The preliminary result showed the efficiency of the model. Currently, the Bernoulli equation, instead of the Navier-Stokes equations is used as the air flow model. To take advantage of the multi-mass model, the Navier-Stokes equations have to be used in the future. Another important issue is how to assign the parameters in the model. A reasonable and systematic method has to be applied to make this model useful.

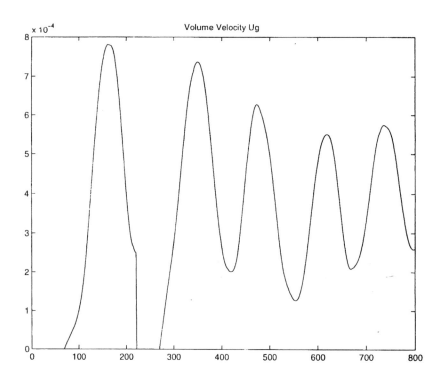

Figure 4.3: Computed glottal wavefrom.

CHAPTER 5

Vocal Fold Excitation Models

In articulatory synthesis, the vocal tract must be excited by some appropriate signal. The choice and method employed to excite the synthesizer play a vital part in determining the spectral characteristics and the degree of naturalness of the synthesized speech. Most speech synthesizers based on LPC techniques assume source-tract separability. For voiced excitations, a sequence of impulses is generated with the spacing determined by the desired fundamental frequency. The unvoiced excitation is generated by a band-limited white noise signal. In order to incorporate source-tract interaction into an articulatory synthesizer, a more realistic model of the glottal excitation must be considered. Two methods for voiced excitation, namely the parametric model and the multimass model of vocal fold excitation are presented in the following sections.

5.1   PARAMETRIC MODELS

Holmes [63], Rothenberg [64], and several other researchers investigated an alternative approach to inverse filtering of the speech waveform to generate the excitation signal. Their findings show that this approach is capable of giving a good estimate of the glottal volume velocity. In the following sections, we will describe two representative parametric models developed by Rosenberg [65] and Titze [66], respectively.

5.1.1   ROSENBERG'S MODEL

In 1971, Rosenberg applied a pitch-synchronous resynthesis method to produce speech utterances with various source waveforms [65]. In his perceptual tests, the most natural excitation signal involves specification of several parameters. In order to better explain Rosenberg's model, we need to first introduce the general waveform of the glottal area during the vocal fold excitation.

As shown in Figure 5.1, T denotes the pitch period, T_P denotes the opening time during the glottal excitation, and T_N denotes the closing time during the glottal excitation. In Rosenberg's model, the glottal waveform is specified by four parameters, namely, amplitude factor α, pitch period T, *open quotient* $\frac{T_P+T_N}{T}$ which denotes the ratio of pulse duration to pitch period, and *speed quotient* $\frac{T_P}{T_N}$ which denotes the ratio of the rising to falling pulse durations. The glottal area function $A_g(t)$ is given by

$$A_g(t) = \begin{cases} \alpha[3(\frac{t}{T_P})^2 - 2(\frac{t}{T_P})^3] & 0 \leq t \leq T_P \\ \alpha\,[1 - (\frac{t-T_P}{T_N})^2] & T_P < t \leq T_P + T_N \\ 0 & T_P + T_N < t \end{cases} \qquad (5.1)$$

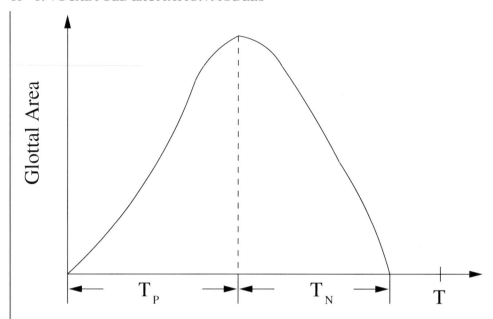

Figure 5.1: General waveform of the glottal area during excitation.

Figure 5.2 shows a glottal waveform computed from Rosenberg's model, where the parameters are set as $\alpha = 1.0$, $T = 10$ ms, $T_P = 0.33T$, and $T_N = 0.1T$.

5.1.2   TITZE'S MODEL

In 1982, Titze proposed a parametric model to represent the glottal area [66]. The glottal waveform in Titze's model is similar to the Rosenberg pulse described in Equation (5.1), with an extra parameter β to determine the residual decay of the falling slope. The glottal area function in this model is given by

$$A_g(\theta) = \begin{cases} \alpha[(\frac{\theta}{\theta_m})^{-\theta_m cot\theta_m}(\frac{sin\theta}{sin\theta_m})]^\beta & \theta \leq \pi \\ 0 & \theta \end{cases} \tag{5.2}$$

where $\theta \triangleq \frac{\pi t}{T_P + T_N}$ and $\theta_m \triangleq \frac{\pi T_P}{T_P + T_N}$.

Figure 5.3 shows a glottal waveform computed from Titze's model, where the parameters are set as $\alpha = 1.0$, $T = 10$ ms, $T_P = 2.4T_N$, $(T_P + T_N) = 0.66T$, and $\beta = 1.2$, respectively.

5.2   MECHANICAL MODEL

Over the years, several researchers have developed a number of different methods for realistic modeling of the vocal fold excitation during speech production. In 1972, Ishizaka and Flanagan [42]

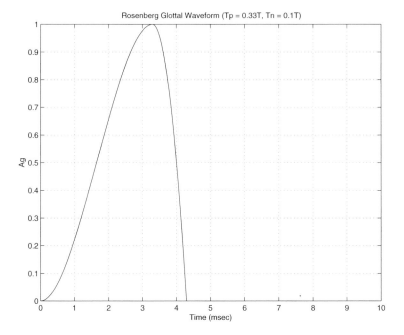

Figure 5.2: Glottal waveform computed from Rosenberg's model.

developed a two-mass model of the vocal folds by choosing the parameters from mechanical considerations and including aerodynamics with estimations of glottal waves produced in normal phonation. Other researchers have proposed similar vocal fold models which differ from each other in their model parameters and means of aerodynamic influences [67], [68]. Recently, researchers have introduced the finite-element method (FEM) to model the detailed geometric and material information in the vocal folds [69], [70]. In this book, we focus on the lumped parameter model of the vocal folds and its application to articulatory synthesis and the experimental evidence of source-tract interaction.

5.2.1 TWO-MASS MODEL

Figure 5.4 shows that the two-mass model describes one vocal fold by two coupled oscillators, where P_s, A_g, and U_g denote the subglottal pressure, glottal area, and glottal volume velocity, respectively. Each oscillator consists of a mass, a spring stiffness, and a damper. Mass m_1, linear spring stiffness k_1, and damper r_1 represent the lower part of the vocal fold with thickness d_1. Mass m_2, linear spring stiffness k_2, and damper r_2 represent the upper part with thickness d_2. The two masses are coupled by a spring stiffness k_c. The two masses m_1 and m_2 are permitted to move in a lateral direction. The deflections of m_1 and m_2 are x_1 and x_2, respectively. In the two-mass model, symmetry along the length of the glottis is assumed; therefore only one vocal fold is considered. When the vocal fold

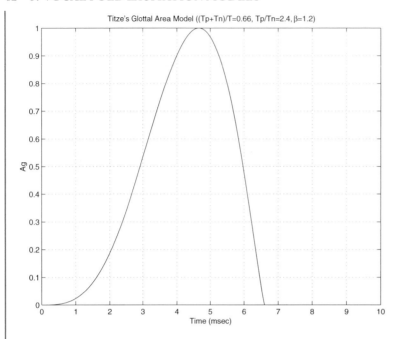

Figure 5.3: Glottal waveform computed from Titze's model.

approaches the symmetry line within a very short distance, collision springs will be activated and have an influence on m_1 and m_2, respectively, in the contralateral direction.

The dynamic response of the mechanical part of the two-mass model can be described by the equations of motion of the two masses as follows:

$$m_1\ddot{x}_1 + r_1\dot{x}_1 + (k_1 + k_c)x_1 - k_c x_2 = F_1 \tag{5.3}$$

$$m_2\ddot{x}_2 + r_2\dot{x}_2 + (k_2 + k_c)x_2 - k_c x_1 = F_2 \tag{5.4}$$

where F_1 and F_2 are the aerodynamic forces exerted on the masses, and $\dot{x}$ and $\ddot{x}$ denote the first and second time derivatives of the variable x, respectively. In our system, an improved two-mass model is used. The improved equations of motion are given as follows:

$$m_1\ddot{x}_1 + r_1\dot{x}_1 + s_1(x_1) + k_c(x_1 - x_2) = F_1 \tag{5.5}$$

$$m_2\ddot{x}_2 + r_2\dot{x}_2 + s_2(x_2) + k_c(x_2 - x_1) = F_2 \tag{5.6}$$

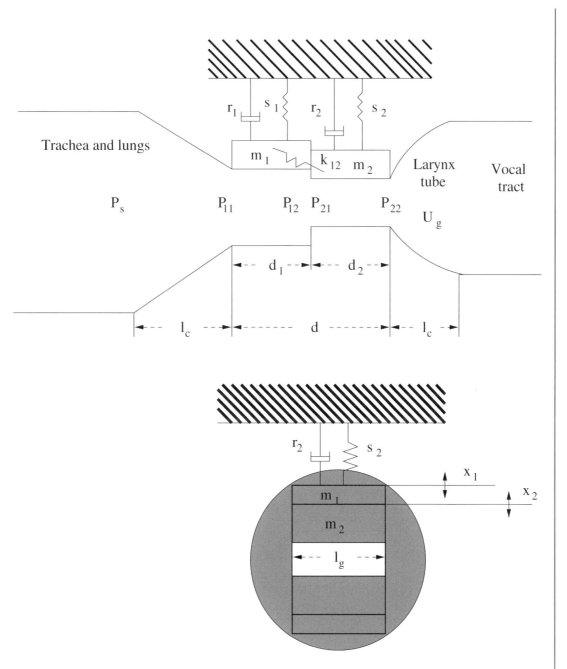

Figure 5.4: Two-mass model of the vocal fold.

$$s_1(x_1) = \begin{cases} k_1(x_1 + \eta_{k1}x_1^3) + h_1[(x_1 + \frac{A_{g01}}{2l_g}) + \eta_{h1}(x_1 + \frac{A_{g01}}{2l_g})^3], & x_1 \leq -\frac{A_{g01}}{2l_g} \\ k_1(x_1 + \eta_{k1}x_1^3), & x_1 > -\frac{A_{g01}}{2l_g} \end{cases} \tag{5.7}$$

$$s_2(x_2) = \begin{cases} k_2(x_2 + \eta_{k2}x_2^3) + h_2[(x_2 + \frac{A_{g02}}{2l_g}) + \eta_{h2}(x_2 + \frac{A_{g02}}{2l_g})^3], & x_2 \leq -\frac{A_{g02}}{2l_g} \\ k_1(x_2 + \eta_{k2}x_2^3), & x_2 > -\frac{A_{g02}}{2l_g} \end{cases} \tag{5.8}$$

where A_{g01} and A_{g02} denote the area through mass 1 and mass 2 at the phonation neutral position, l_g denotes the depth of glottis slit, k_1 and k_2 denote the open linear spring constant of mass 1 and mass 2, h_1 and h_2 denote the closed linear spring constant of mass 1 and mass 2, η_{k1} and η_{k2} denote the open nonlinear spring constant of mass 1 and mass 2, and η_{h1} and η_{h2} denote the closed nonlinear spring constant of mass 1 and mass 2.

The vocal fold model is coupled with the aerodynamic equations by relating the aerodynamic forces F_1 and F_2 to the glottal volume velocity U_g in the following way:

$$P_s - P_{11} = 1.37\frac{\rho}{2}(\frac{U_g}{A_{g1}})^2 + \int_0^{l_c} \frac{\rho}{A_c(x)}dx\frac{dU_g}{dt} \tag{5.9}$$

$$P_{11} - P_{12} = 12\frac{\mu l_g^2 d_1}{A_{g1}^3}U_g + \frac{\rho d_1}{A_{g1}}\frac{dU_g}{dt} \tag{5.10}$$

$$P_{12} - P_{21} = \frac{\rho}{2}U_g^2(\frac{1}{A_{g2}^2} - \frac{1}{A_{g1}^2}) \tag{5.11}$$

$$P_{21} - P_{22} = \frac{\mu l_g^2 d_2}{A_{g2}^3}U_g + \frac{\rho d_2}{A_{g2}}\frac{dU_g}{dt} \tag{5.12}$$

$$P_{22} - P_1 = -\frac{\rho}{2}(\frac{U_g}{A_{g2}})^2\frac{A_{g2}}{A_1}(1 - \frac{A_{g2}}{A_1}) \tag{5.13}$$

$$F_1 = \frac{1}{2}(P_{11} + P_{12})l_g d_1 \tag{5.14}$$

$$F_2 = \frac{1}{2}(P_{21} + P_{22})l_g d_2 \tag{5.15}$$

where $A_c(x)$ denotes the contraction area at x, A_1 denotes the area after expansion, and A_{g1} and A_{g2} denote the area through mass 1 and mass 2, respectively.

The numerical solutions of the vocal fold model equations are computed in the following iterative way. In the nth iteration, the volume velocity computed from the previous iteration $U_g^{(n-1)}$ is used to compute the aerodynamic forces $F_1^{(n)}$ and $F_2^{(n)}$ using Equations (5.9) - (5.15). Then the computed forces $F_1^{(n)}$ and $F_2^{(n)}$ are substituted into Equations (5.5) - (5.8) to compute the mass displacements $x_1^{(n)}$ and $x_2^{(n)}$. From $x_1^{(n)}$ and $x_2^{(n)}$, we further compute the areas $A_{g1}^{(n)}$ and $A_{g2}^{(n)}$. Finally, the glottal area in the nth iteration is determined by $A_g^{(n)} = min\{A_{g1}^{(n)}, A_{g2}^{(n)}\}$. This glottal area is used to compute the volume velocity for next iteration, which is used as one of the boundary conditions for the Navier-Stokes equations during the next iteration.

5.2.2   M-MASS MODEL

In the current synthesis system, the two-mass vocal fold model described in the previous section is used to compute the glottal excitation signals for articulatory speech synthesis. In reality, an M-mass model has the potential to compute more accurate glottal excitation signals for articulatory speech synthesis.

The equations of motion for the M-mass model as follows:

$$
\begin{aligned}
&m_1\ddot{x}_1 + r_1\dot{x}_1 + k_1x_1 + k_{1,2}(x_1 - x_2) = F_1 \\
&m_i\ddot{x}_i + r_i\dot{x}_i + k_ix_i + k_{i,i-1}(x_i - x_{i-1}) + k_{i,i+1}(x_i - x_{i+1}) = F_i, \quad i = 2, \cdots, M-1 \\
&m_M\ddot{x}_M + r_M\dot{x}_M + k_Mx_M + k_{M,M-1}(x_M - x_{M-1}) = F_M
\end{aligned} \tag{5.16}
$$

where k_i denotes the spring stiffness of mass m_i, r_i denotes the damping coefficient of mass m_i, and k_{ij} denotes the coupling spring stiffness between mass m_i and mass m_j. Alternatively, we can write the equations of motion of the M-mass model in matrix form as

$$
M_{mass}\ddot{\vec{x}} + R\dot{\vec{x}} + K\vec{x} = \vec{F} \tag{5.17}
$$

where M_{mass}, R, and K denote the mass matrix, damping matrix, and stiffness matrix, respectively. However, if we want to consider the nonlinear effects and both the open spring and closed spring constants similar to the improved two-mass model described in the previous section, the simple matrix form in Equation (5.17) cannot be used and terms k_ix_i, $i = 1, \cdots, M$ in Equation (5.16) should be replaced by $s_i(x_i)$. The $s_i(x_i)$ are given by

$$
s_i(x_i) = \begin{cases} k_i(x_i + \eta_{ki}x_i^3) + h_i[(x_i + \frac{A_{g0i}}{2l_g}) + \eta_{hi}(x_i + \frac{A_{g0i}}{2l_g})^3], & x_i \leq -\frac{A_{g0i}}{2l_g} \\ k_i(x_i + \eta_{ki}x_i^3), & x_i > -\frac{A_{g0i}}{2l_g} \end{cases} \tag{5.18}
$$

where A_{g0i} denotes the area through mass i at the phonation neutral position, k_i denotes the open linear spring constant of mass i, h_i denotes the closed linear spring constant of mass i, η_{ki} denotes the open nonlinear spring constant of mass i, and η_{hi} denotes the closed nonlinear spring constant of mass i, for $i = 1, \cdots, M$.

5.3   SIMULATION RESULTS

In this section, we present the simulation results for vocal excitation based on the M-mass model, as described above, with $M = 2$ described above. Our experimental results include the glottal area

A_g, glottal particle velocity V_g, and glottal volume velocity U_g of the English sentence *Where were you while we were away?* The glottal particle velocity V_g and glottal volume velocity U_g will be used as initial conditions for the Navier-Stokes (N-S) equations for articulatory synthesis which will be discussed in the next chapter. The different experimental conditions are listed in Table 5.1.

Table 5.1: Experimental conditions for the simulation of glottal excitation.

ID	VTLN	Time-varying pitch	Duration of sentence
Run 1	Yes (factor = 1.4)	No	2.4 sec
Run 2	No	No	2.4 sec
Run 3	No	Yes	2.4 sec
Run 4	No	Yes	1.2 sec

In Table 5.1, VTLN denotes vocal tract length normalization. The reason for using VTLN is to compensate for the unrealistic vocal tract length (greater than 23 cm in most cases) computed from Coker's model. In the case of time-invarying pitch, a single pitch ($F_0 = 100$ Hz) is used throughout the simulation. In the case of time-varying pitch, the pitch contour of a waveform of the sentence *Where were you while we were away?* from the TIMIT database is extracted using the ESPS/Xwaves+ tool and applied to the glottal excitation simulation. The duration of the original sentence from TIMIT database is 1.2 s. In our experiments, we found that the speaking rate of this sentence was too fast to synthesize a natural sounding speech signal, so we artificially doubled the synthesis length in order to compensate for the flaw in the database. We have also segmented the phonetic boundaries of several recorded normal speaking rate utterances and will use this prosodic information to resynthesize continuous speech sentences later. The waveforms of the simulated glottal area A_g, glottal particle velocity V_g, and glottal volume velocity U_g are shown in Figures 5.5 - 5.8, Figures 5.9 - 5.12, and Figures 5.13 - 5.16, respectively. In Figures 5.5 - 5.16, one time step corresponds to 10^{-5} s.

Finally, we compare the pitch contour of the original speech from the TIMIT database and the synthesized speech from Run 1. The resulting pitch contour plot is shown in Figure 5.17. From this figure, we can see that the general trend of pitch contour of the synthesized speech from Run 1 is very similar to that of the recorded speech from TIMIT. Since we did not change the spring stiffness constants during the Run 1 simulation, the variation of the pitch contour is purely due to the interaction between the vocal tract and the vocal fold. This means that there are not only flow parts of the source-tract interaction as shown in the ripples of the glottal particle velocity and glottal volume velocity in Figures 5.9 - 5.16, but also a mechanical part of the source-tract interaction as shown in the change of the pitch contour in Figure 5.17. In order to prove this hypothesis, we plot the glottal area from Run 1 at each 2-ms time step in Figure 5.18. We can see clear variation of the outline contour of the glottal area which is due to the mechanical part of the source-tract interaction. This is a very interesting source-tract interaction phenomenon, which to our knowledge, has never been observed before.

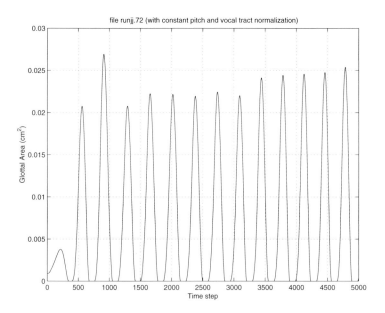

Figure 5.5: Glottal area—Run 1.

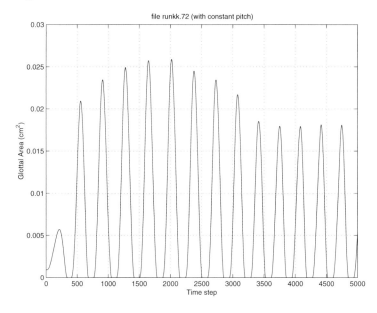

Figure 5.6: Glottal area—Run 2.

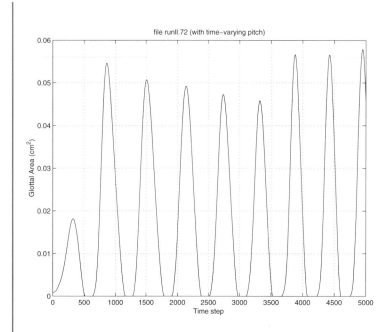

Figure 5.7: Glottal area—Run 3.

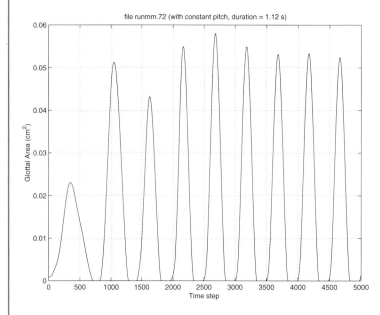

Figure 5.8: Glottal area—Run 4.

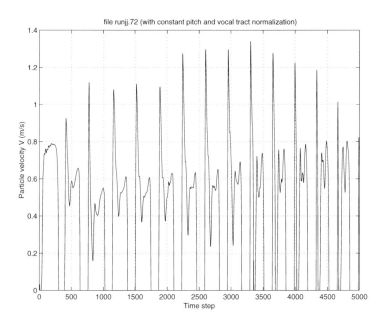

Figure 5.9: Glottal particle velocity—Run 1.

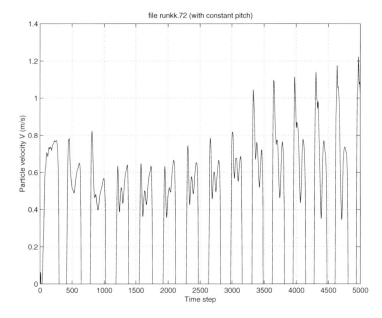

Figure 5.10: Glottal particle velocity—Run 2.

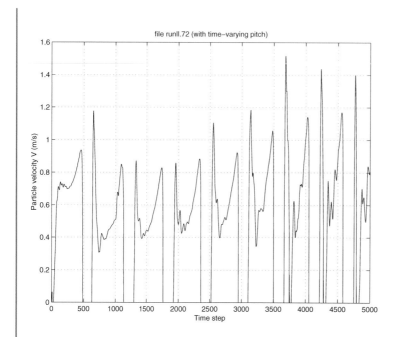

Figure 5.11: Glottal particle velocity—Run 3.

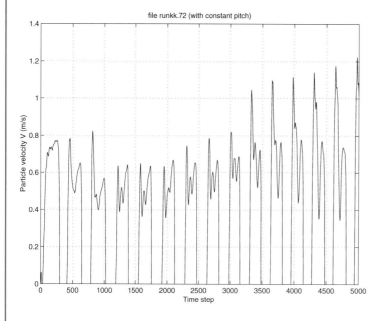

Figure 5.12: Glottal particle velocity—Run 4.

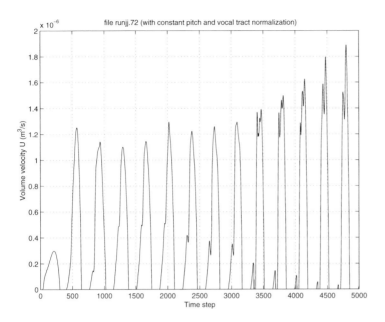

Figure 5.13: Glottal volume velocity—Run 1.

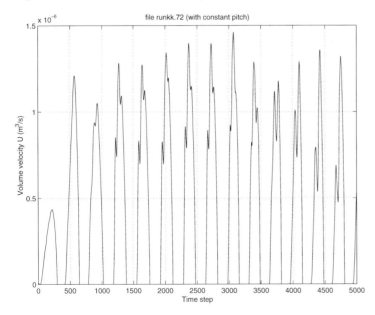

Figure 5.14: Glottal volume velocity—Run 2.

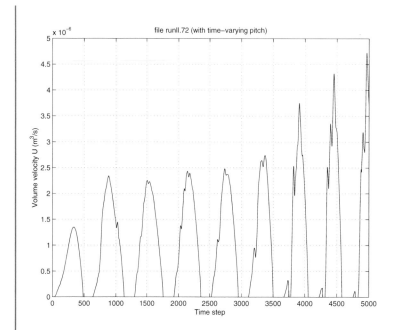

Figure 5.15: Glottal volume velocity—Run 3.

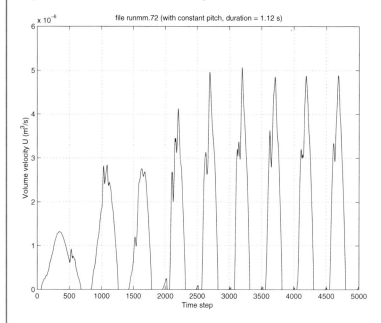

Figure 5.16: Glottal volume velocity—Run 4.

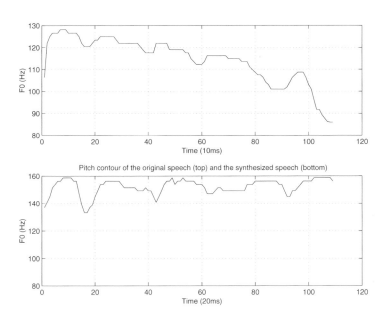

Figure 5.17: Pitch contour of the recorded speech and the synthesized speech (Run 1) of the sentence *Where were you while we were away?*

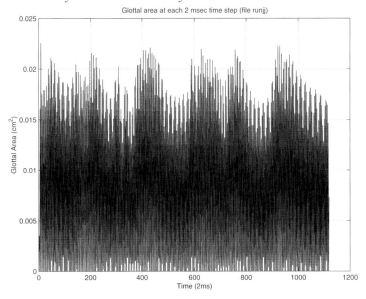

Figure 5.18: Glottal area of the synthesized speech (Run 1) at each 2-ms time step.

5.4   DISCUSSION

From Figures 5.5 - 5.16, we can see that the M-mass model of the vocal folds is able to generate more realistic results for the glottal area and glottal volume velocity than a parametric flow waveform model. This method is superior to the parametric models described in Section 5.1 because we can clearly observe ripples, especially in the positive glottal opening interval of the simulated U_g waveforms, which is strong evidence of the source-tract interaction. Figure 5.19 shows the Fourier analysis of the computed volume velocity for Run 3. From this figure, we can clearly see the influence of the first formant ($F1$) on the source excitation signal.

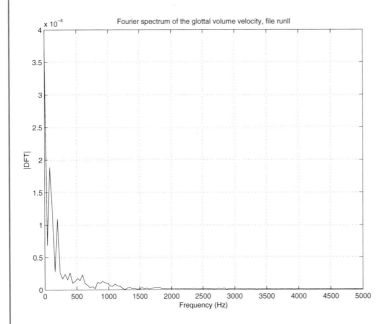

Figure 5.19: Fourier analysis of the volume velocity—Run 3.

Further evidence of source-tract interaction is the observation of the skewing effect of the glottal waveform due to the existence of vocal tract constriction. Our results are similar to those reported by Fant et al. [24], [71], [72]. However, our results also simulate the glottal particle velocity, which is not done in Fant's work. Furthermore, multiple ripples can be observed in our simulation results of glottal particle velocity and glottal volume velocity. Finally, we observed significant pitch contour variation of the synthesized speech from Figure 5.17. This means that the supralaryngeal wave motion due to the change of the vocal tract shape affects the vibration of the vocal folds and subsequently the fundamental frequency of the produced speech. This is new evidence of source-tract interaction that has not been observed before as far as we know. The source-tract separability is unsuitable during the production of unvoiced sounds due to the additional excitation source required at the vocal tract constriction. Even during the production of voiced sounds, this assumption only holds as a first-order approximation since the glottal excitation signal depends on the transglottal pressure which, in turn, is controlled by the subglottal pressure and the vocal tract shape. A more realistic model of the vocal fold excitation, which considers the nonlinear effects due to source-tract interaction, is not only important for speech synthesis, but also beneficial for low-bit rate perceptual speech coding.

5.5   SUMMARY

In this chapter, we presented vocal fold excitation models and some simulation results. First, we briefly discussed the parametric models of vocal fold excitation. Then, we presented our mechanical model of vocal fold excitation, which considers the masses of each part of the vocal fold, damping coefficients of each mass, and the linear and nonlinear spring stiffness coefficients of each mass. In our articulatory synthesizer, the vocal fold and vocal tract were coupled through aerodynamics. Simulation results showed that both the flow part and the mechanical part of the source-tract interaction can be observed in the simulated glottal particle velocity and glottal volume velocity.

CHAPTER 6

Experimental Results of Articulatory Synthesis

6.1 GOVERNING EQUATIONS, FLUID DYNAMICS ANALYSIS

The governing equations of fluid dynamics consist of the continuity, momentum, and energy equations. These equations may be cast in conservation form as:

Continuity

$$\frac{\partial \rho}{\partial t} + \frac{\partial \rho u_i}{\partial x_i} = 0 \tag{6.1}$$

Momentum

$$\frac{\partial \rho u_i}{\partial t} + \frac{\partial \rho u_i u_j}{\partial x_j} + \frac{\partial p}{\partial x_i} - \frac{\partial \tau_{ij}}{\partial x_j} - \rho g_i = 0 \tag{6.2}$$

Energy

$$\frac{\partial \rho e_t}{\partial t} + \frac{\partial \left(\rho u_j e_t + u_j p \right)}{\partial x_j} - \frac{\partial q_i}{\partial x_i} - \frac{\partial u_i \tau_{ij}}{\partial x_j} - \rho g_i u_i = 0 \tag{6.3}$$

In these equations, ρ is density, u_i is the velocity vector, t is time, x_i is the space vector, p is pressure, τ_{ij} is the viscous stress tensor, e_t is the total energy (internal energy + kinetic energy), q_i is the heat conduction vector, and g_i is the gravity vector. In the current work, the flow in the vocal apparatus is assumed adiabatic such that the energy equation need not be solved, and the effects of gravity can be ignored.

Flow velocities approach a maximum of approximately 40 m/s in the vocal apparatus. At these flow velocities, the Mach number (Ma) is relatively low (Ma=0.12) such that spatial gradients of density in the governing equations may be ignored. Neglecting the spatial gradients of density, and assuming an isentropic relationship between density and pressure (dp/dρ =c^2, where c is the speed of sound), Equations (6.1) and (6.2) can be recast in a non-dimensionalized "slightly compressible" form as:

$$Ma^2 \frac{\partial \tilde{p}}{\partial \tilde{t}} + \frac{\partial \tilde{u}_i}{\partial \tilde{x}_i} = 0 \tag{6.4}$$

$$\frac{\partial \tilde{u}_i}{\partial \tilde{t}} + \frac{\partial \tilde{u}_i \tilde{u}_j}{\partial \tilde{x}_j} + \frac{\partial \tilde{p}}{\partial \tilde{x}_i} - \frac{1}{Re} \frac{\partial \tilde{\tau}_{ij}}{\partial \tilde{x}_j} = 0 \tag{6.5}$$

These equations have been non-dimensionalized using reference values of density (ρ_0), velocity (U_0), length scale (L_0), and kinematic viscosity (ν_0), where:

$$\rho = \tilde{\rho}\rho_0, \ u_i = \tilde{u}_i U_0, p = \tilde{p}\rho U_0^2, \ t = \tilde{t}\frac{L_0}{U_0}, \ x_i = \tilde{x}_i L_0,$$

$$\tau_{ij} = \tilde{\tau}_{ij}\frac{U_0 \nu_0}{L_0}, \nu = \tilde{\nu}\nu_0, \ Ma = \frac{U_0}{c}, \ \text{and} \ Re = \frac{U_0 L_0}{\nu_0}$$

Equations (6.4) and (6.5) can be cast as an inhomogeneous wave equation in pressure by subtracting the divergence of Equation (6.5) from the time derivative of Equation (6.4):

$$Ma^2 \frac{\partial^2 \tilde{p}}{\partial \tilde{t}^2} - \frac{\partial^2 \tilde{p}}{\partial \tilde{x}_i \partial \tilde{x}_i} = \frac{\partial^2 \tilde{u}_i \tilde{u}_j}{\partial \tilde{x}_i \partial \tilde{x}_j} - \frac{1}{Re}\frac{\partial^2 \tilde{\tau}_{ij}}{\partial \tilde{x}_i \partial \tilde{x}_j} \tag{6.6}$$

This equation contrasts with the homogeneous wave equation used in traditional speech synthesis, which is based on the assumption of linear plane wave propagation inside the vocal apparatus. The right-hand side terms of Equation (6.6), which are ignored in traditional speech synthesis, include spatial gradients of non-linear convection terms and the viscous stress tensor. The implicit inclusion of these terms in Equations (6.4) and (6.5) provides a solution approach that allows for a more complete physical description of the flow in the vocal apparatus.

At moderate to high Reynolds numbers, Equations (6.4) and (6.5) become computationally intensive to solve as a result of the small length and time scales that must be resolved in turbulent flow simulations. The current state-of-the-practice is to Reynolds average the governing equations. With this approach, the flow variables are separated into mean and fluctuating components and the resulting equations are time averaged. The result is a new set of equations called the Reynolds-averaged Navier Stokes (RANS) equations. These equations are given by:

$$Ma^2 \frac{\partial \bar{p}}{\partial t} + \frac{\partial \bar{u}_i}{\partial x_i} = 0 \tag{6.7}$$

$$\frac{\partial \bar{u}_i}{\partial t} + \frac{\partial \bar{u}_i \bar{u}_j}{\partial x_j} + \frac{\partial \bar{p}}{\partial x_i} - \frac{1}{Re}\frac{\partial \bar{\tau}_{ij}}{\partial x_j} = 0 \tag{6.8}$$

where the overbar indicates a time-averaged quantity and the tilde's have been dropped for convenience. The last term on the left-hand side of the time-averaged momentum equations may be written as:

$$-\frac{1}{Re}\frac{\partial \bar{\tau}_{ij}}{\partial x_j} = -\frac{1}{Re}\frac{\partial}{\partial x_j}\left(\nu\left[\frac{\partial \bar{u}_i}{\partial x_j} + \frac{\partial \bar{u}_j}{\partial x_i}\right]\right) + \frac{\partial}{\partial x_j}\left(\overline{u_i' u_j'}\right) \tag{6.9}$$

where the primed variables represent fluctuating components. In this form, the stress tensor has been expressed in terms of velocity gradients and a second term that contains time averages of products of fluctuating velocities has been added. The elements of the added term are called turbulent or Reynolds stresses. Using the Boussinesq assumption, which relates the turbulent stresses to the mean flow variables, the last term in Equation (6.9) can be written as:

$$\frac{\partial}{\partial x_j}\left(\overline{u_i' u_j'}\right) = -\frac{1}{Re}\frac{\partial}{\partial x_j}\left(\nu_t\left[\frac{\partial \bar{u}_i}{\partial x_j} + \frac{\partial \bar{u}_j}{\partial x_i}\right]\right) \tag{6.10}$$

where ν_t represents the turbulent or eddy viscosity. Using Equations (6.9) and (6.10), Equations (6.7) and (6.8) can be rewritten as:

$$Ma^2 \frac{\partial \bar{p}}{\partial t} + \frac{\partial \bar{u}_i}{\partial x_i} = 0 \qquad (6.11)$$

$$\frac{\partial \bar{u}_i}{\partial t} + \frac{\partial \bar{u}_i \bar{u}_j}{\partial x_j} + \frac{\partial \bar{p}}{\partial x_i} - \frac{1}{Re} \frac{\partial}{\partial x_j} \left((\nu + \nu_t) \left[\frac{\partial \bar{u}_i}{\partial x_j} + \frac{\partial \bar{u}_j}{\partial x_i} \right] \right) = 0 \qquad (6.12)$$

Since ν_t represents an additional unknown, one or more additional equations, called turbulence models, must be included to account for it.

A wide range of turbulence models of varying degrees of complexity exist. Given the inherent difficulties (i.e., complex geometry, moving grids, complex time-varying flows) associated with solving for the turbulent flow in the vocal apparatus, it was decided that using a simple algebraic turbulence model made the most sense. It was felt that this model would provide an acceptable level of accuracy, would provide much needed efficiencies, and would mitigate unnecessary computational difficulties. The turbulence model that was used is given by:

$$\nu_t = \ell^2 \left(2\bar{S}_{ij} \bar{S}_{ij} \right)^{1/2} \qquad (6.13)$$

where:

$$\bar{S}_{ij} = \frac{1}{2} \left(\frac{\partial \bar{u}_i}{\partial x_j} + \frac{\partial \bar{u}_j}{\partial x_i} \right) \qquad (6.14)$$

and:

$$\ell = D\ell_\infty \tanh (\kappa d / \ell_\infty) \qquad (6.15)$$

In Equation (6.15), ℓ_∞ is a specified outer-region value of mixing length, d is the distance from the wall, κ is the von Karman constant, and D is the van Driest damping factor.

For our studies, Equations (6.11) and (6.12), along with the eddy viscosity model given by Equation (6.13), were solved using a proprietary RANS-based flow solver developed at Electric Boat Corporation. This solver is finite difference-based and uses a Briley-McDonald linearized block, alternating direction implicit scheme. The governing system of conservation equations is transformed to generalized curvilinear coordinates and, for the applications presented in this book, is discretized using central spatial differencing, with full non-orthogonal viscous terms. A hybrid second and fourth order, implicit and explicit artificial dissipation scheme is used to enhance solution stability. All artificial dissipation is applied locally and is adjusted each time step based on nodal checks of grid quality and solution stability obtained from grid cell Reynolds number and local eigenvalues. This approach establishes dissipation levels for optimum solution convergence behavior and quality.

To accommodate the movement of the vocal tract, time-dependent metrics are included. For these terms, wall velocities (i.e., the velocity at each computational node on the vocal tract walls) are computed at each time step from an articulatory model. At the inlet to the vocal tract, the RANS solver is coupled to the multimass vocal fold model through boundary conditions. At each time step, the RANS solver uses velocities computed from the multimass model as upstream boundary conditions while the pressure at the vocal tract inlet is extrapolated from the interior. In turn, the multimass model uses the pressure computed during the RANS solution as a boundary condition for the next time step.

6.2   SYNTHESIZED WAVEFORM

In this section, we present computational results from our articulatory speech synthesis system. The simulation time step for the RANS-solver is 10^{-5} s. The vocal fold model described in Chapter 5 is coupled with the RANS-solver and they are iteratively solved. First, we present the synthesized speech waveform of a dipthong /AY/ ("uy" in buy). Forty vocal tract shape were estimated using the dynamic parameter estimation method described in Chapter 3. The time step between each consecutive vocal tract shape is 10 ms. Moving grids, i.e., dynamic metrics terms, were added to the RANS solver to accurately account for the actual movement. The boundary condition at the vocal tract inlet is specified by the particle velocity computed from the vocal fold excitation model. Figure 6.1 shows the waveform of the synthesized dipthong /AY/.

In the next step, we synthesize the speech waveforms of the English sentence *Where were you while we were away?* The prosodic information (pitch contour, duration, and short-time energy) were extracted from a sentence in the TIMIT database. Different simulation conditions were listed in Table 5.1. Figures 6.2 - 6.5 show the waveforms of the synthesized speech sentence for Run 1, Run 2, Run 3, and Run 4.

6.3   SPEECH ANALYSIS RESULTS

6.3.1   LPC SPECTRUM AND THE SHORT-TIME POWER SPECTRUM

In this section, the LPC spectrum and the short-time power spectrum of the synthesized speech signal are computed. A preemphasis filter with the parameter $\alpha = 0.97$ is used to equalize the inherent spectral tilt in speech. A Hamming window with the length 24 ms is applied for preprocessing. The order of the LPC analysis is 12. The short-time power spectrum is computed through the FFT technique. Figure 6.6 shows the spectra of the second frame of the synthesized dipthong /AY/ (*uy* in *buy*, which corresponds to the spectra of vowel /AH/ (*u* in *but*). The upper figure denotes the LPC spectrum and the lower figure denotes the short-time power spectrum, respectively. The sampling frequency is 20 kHz.

6.3.2   SPECTROGRAM

In this section, we present the spectrograms of the synthesized sentence *Where were you while we were away?* The same preemphasis and Hamming windowing approach as described in the previous section are applied. The window length is 25.6 ms and a 50% overlapping is applied before the time-frequency analysis. We need to note that the static articulatory parameters were based on Levinson and Schmidt's [31] acoustic-to-articulatory mapping algorithm. The speakers of the speech signals

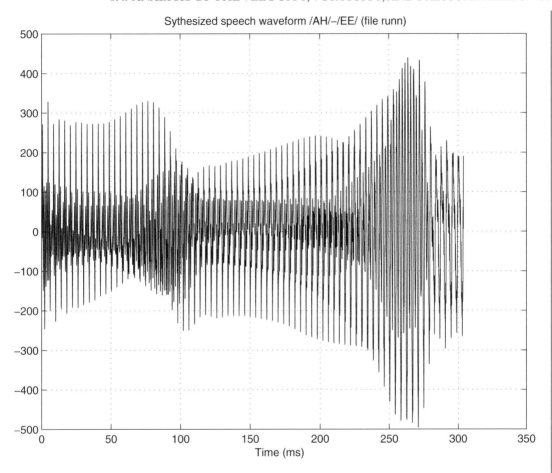

Figure 6.1: Waveform of the synthesized dipthong /AY/ ("uy" in buy).

for static articulatory parameter estimation are different from the speakers in the TIMIT database. We use the speech data from the TIMIT database only for the purpose of prosodic information (phoneme duration and pitch contour) extraction.

6.4   ANALYSIS OF THE VELOCITY, VORTICITY, AND PRESSURE FIELDS

Figures 6.7 - 6.12 show the snapshot plots of the vector velocity fields at times 0.2 ms, 2 ms, 4 ms, 6 ms, 7.5 ms, and 10 ms, respectively. We can see from these figures that there is a lot of fluid dynamic activity during the production of human speech. At the beginning of phonation, high-speed flow was formed at the glottal exit, as shown in Figure 6.7. When the flow encounters a sudden expansion in

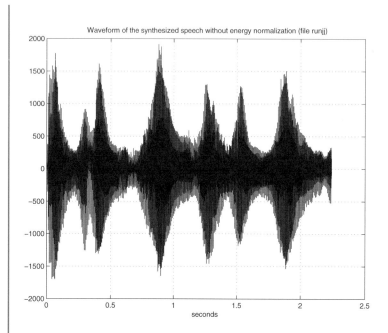

Figure 6.2: Waveform of the synthesized sentence *Where were you while we were away?* for Run 1.

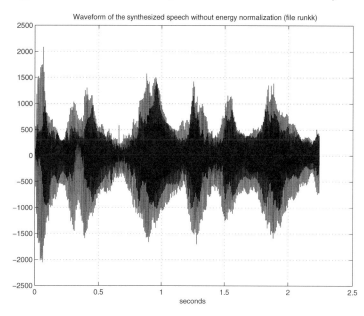

Figure 6.3: Waveform of the synthesized sentence *Where were you while we were away?* for Run 2.

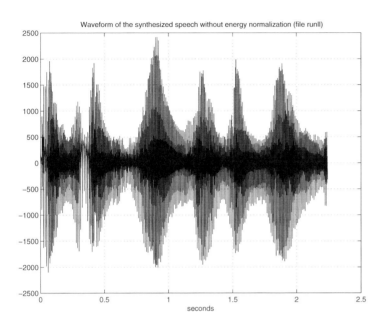

Figure 6.4: Waveform of the synthesized sentence *Where were you while we were away?* for Run 3.

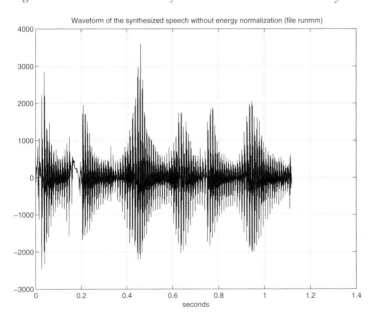

Figure 6.5: Waveform of the synthesized sentence *Where were you while we were away?* for Run 4.

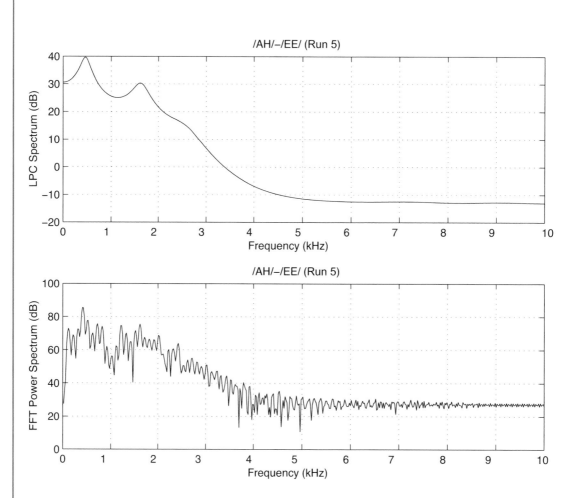

Figure 6.6: LPC spectrum and short-time power spectrum of the synthesized dipthong /AY/ ("uy" in buy).

the vocal tract, high-speed flow separates from the relatively stagnant fluid on the downstream side of the expansion wall, resulting in a turbulent jet. A turbulent jet is characterized by irregular, high-frequency fluctuation in velocity at a given point in space. The jet then mixes with the surrounding air as shown in Figure 6.8, which corresponds to the mixing stage. As the jet travels along the vocal tract, it will split into $2 - 3$ parts. The boundary between each region has the characteristic of shear layer as shown in the zoom-out plot of a portion of the velocity field at 7.5 ms in Figure 6.13. Finally, the fully developed jet gradually vanishes while expanding along the vocal tract as shown in Figure 6.12. A similar pattern repeats itself during the next cycle of vocal fold vibration.

Figures 6.14 - 6.17 show the contour plot of the vorticity fields at the times of 2 ms and 8 ms, and the pressure fields at the time of 0.2 ms and 5 ms, respectively. We can see from Figures 6.14 - 6.15 that the vorticity originally exists in the vocal tract wall boundary layer as shown in Figure 6.14. As the jet travels along the vocal tract, vorticity, such as that shown by the red patches in Figure 6.15, will form inside the vocal tract. When the vorticity propagates along the vocal tract, its range gradually increases and its strength decreases. We can also see from Figures 6.16 - 6.17 that the pressure field inside the vocal tract changes along with the velocity and vorticity fields.

6.5   SUMMARY

In this chapter, we presented experimental results on our articulatory speech synthesizer. First, we presented the waveforms of synthesized dipthong and continuous speech sentences. Second, we showed the spectral analysis which includes the LPC spectrum and the spectrograms of the synthesized speech. Finally, we presented several snapshots of the simulation results of vector velocity fields, vorticity fields, and pressure fields of a part of the dipthong /AY/. We can see from the simulation results that wave propagation inside the human vocal apparatus includes both irrotational and rotational flow components.

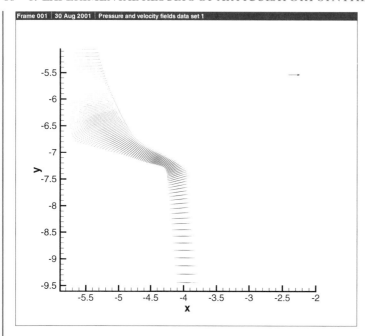

Figure 6.7: Velocity field at the time of 0.2 ms.

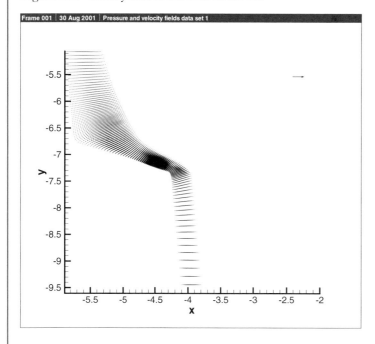

Figure 6.8: Velocity field at the time of 2 ms.

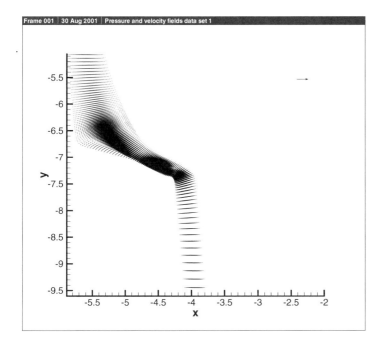

Figure 6.9: Velocity field at the time of 4 ms.

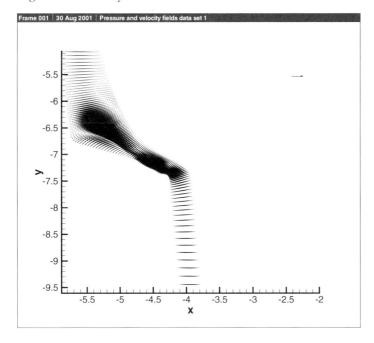

Figure 6.10: Velocity field at the time of 6 ms.

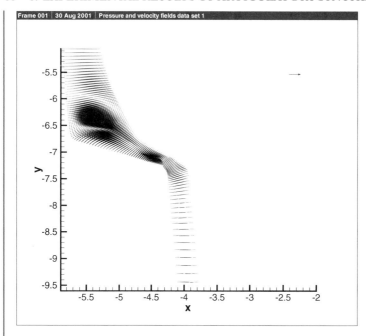

Figure 6.11: Velocity field at the time of 7.5 ms.

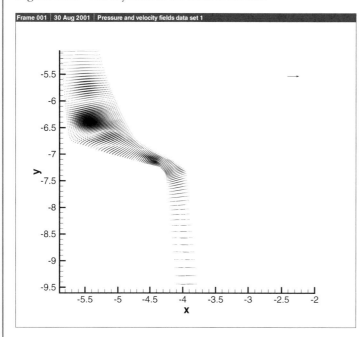

Figure 6.12: Velocity field at the time of 10 ms.

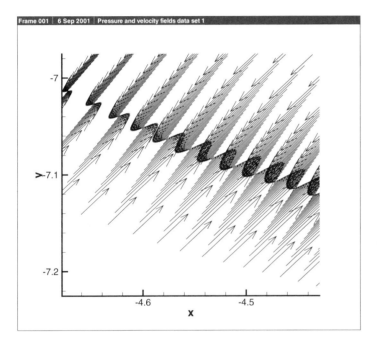

Figure 6.13: Zoom-out plot of a portion of the velocity field at the time of 7.5 ms.

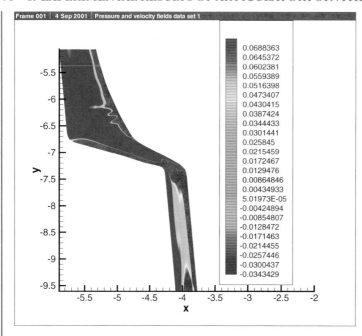

Figure 6.14: Contour plot of the vorticity field at the time of 2 ms.

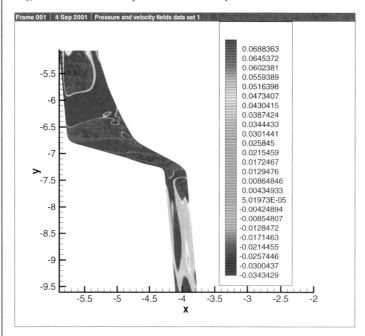

Figure 6.15: Contour plot of the vorticity field at the time of 8 ms.

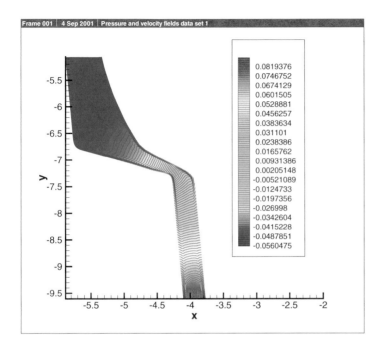

Figure 6.16: Contour plot of the pressure field at the time of 2 ms.

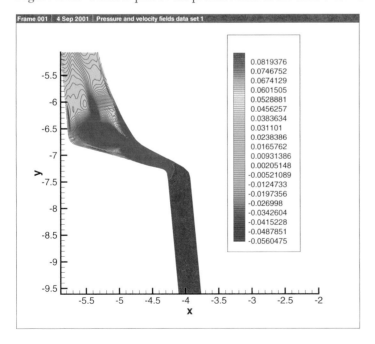

Figure 6.17: Contour plot of the pressure field at the time of 8 ms.

CHAPTER 7

Conclusion

In this book, a complete framework for an articulatory speech synthesizer and its relation to speech production is presented. A combined minimum square error and minimum jerk criterion is proposed to solve the motor control problem in speech production. This approach is proven to avoid the overshooting problems in the interpolation-based methods, and it is successfully applied to estimate the moving vocal tract shape in articulatory synthesis. A vocal fold model based on fundamentals of mechanics and fluid mechanics is used to compute the excitation signal for the vocal tract. The advantage of this model is that it closely couples the vocal fold and vocal tract systems. As a result, we can obtain a vocal fold excitation signal with much more detail than the smooth signal obtained from the conventional parametric models of vocal folds. Strong evidence of source-tract interaction is also found in the simulation results. Furthermore, an articulatory synthesis system was implemented to synthesize intelligible phonemes and continuous speech sentences. Our finding is that articulatory synthesis has the potential to synthesize human-like speech sounds and has more flexibility than conventional approaches. For example, articulatory synthesis uses interpolation methods naturally related to articulator movement planning to cope with the coarticulation and unit concatenation problem. The vocal fold excitation is naturally computed from a mechanical model and the aeroacoustics. On the other hand, smoothing between the unit boundaries is still a major problem in concatenative synthesis, and an accurate model to represent the excitation signal is still not completely found in formant synthesis. Furthermore, the articulatory synthesizer is able to conveniently synthesize personalized speech sounds. Finally, we propose a computational model of the fricative and stop types of sounds based on the physical principles of speech production. The advantage of this model is that it considers the additional nonstationary and nonlinear effects due to the flow mode which are ignored by the conventional source-filter speech production model. A recursive algorithm is used to estimate the model parameters. Experimental results show that this model is able to synthesize natural quality fricative and stop types of sounds which are very close to human speech.

During the last decades, speech signal processing has undergone significant improvements along with progress in the areas of digital signal processing, applied statistics, VLSI design, pattern recognition, artificial intelligence, etc. However, the speech and language problem is far from being completely solved. In [78], Rabiner and Levinson proposed an alternative view of the speech production/speech perception process. From their view, the discrete symbol information rate in the raw message text is about 50 bits per second (bps), the information rate after the language code conversation and inclusion of prosody is about 200 bps, the information rate at the neuromuscular level is about 2000 bps, and finally the information rate of the acoustic signal is between 30,000 and

50,000 bps. On the other hand, we can clearly see that the current speech recognition, synthesis, and coding techniques use a much higher rate to process the speech signal while their performances are still much worse than those of the human being. The articulatory speech production model has the potential to lead to a more compact model of the speech signal and thus to be beneficial to speech signal processing. Although limited research has been conducted in this area compared to other speech processing techniques, we can cautiously claim that investigation of speech production based on first physical principles - and with the help from cross-disciplinary research including signal processing, statistics, fluid dynamics, aeroacoustics, and aerodynamics - might help us to solve some bottleneck problems encountered in current speech signal processing.

Although articulatory synthesis and speech production modeling has the potential to help solve a wide range of problems in speech signal processing, we realize that a lot of research remains to be conducted to address the unsolved problems, computational complexity, and even performance of the technique. Finally, some interesting research topics are as follows:

1. Investigation of the effects of nonsymmetric, moving, compliant walls of the vocal tract on speech generation.
2. A much more simplified but still effective computational model which considers both the convective and the propagative modes during speech production.
3. Inclusion of the nasal tract in an articulatory model.
4. Investigation of relations between different speaking styles such as stress, accent, and articulatory parameters, and applications to personalized speech synthesis.
5. Investigation of the solution of the RANS equations when there is a closure of the vocal tract.
6. Investigation of the solution of the RANS equations when there is a closure of the vocal cords.
7. Inclusion of the shear force due to fluid dynamics.
8. Construction of the articulatory model based on MRI data.

Bibliography

[1] R. Paget, *Human Speech: Some Observations, Experiments and Conclusions as to the Nature, Origin, Purpose and Possible Improvement of Human Speech.* London, UK: Harcourt, 1920. 1

[2] J. L. Flanagan, *Speech Analysis, Synthesis, and Perception.* New York: Springer-Verlag, 2nd edition, 1972. 1, 15, 53, 54

[3] H. Dudley, R. R. Riesz, and S. A. Watkins, "A synthtic speaker," *J. Franklin Inst.*, vol. 227, 1939, pp. 739-764. DOI: 10.1016/S0016-0032(39)90816-1 1

[4] H. Dudley, "The vocoder," *Bell Laboratories Record*, vol. 17, pp. 122-126, 1939. 2

[5] A. M. Liberman, F. Ingeman, L. Lisker, P. Delattre, and F. S. Cooper, "Minimal rules for synthesising speech," *J. Acoust. Soc. Am.*, vol. 31, 1959, pp. 1490-1499. DOI: 10.1121/1.1907654 2

[6] J. L. Kelley and L. J. Gerstman, "An artificial talker driven from phonetic input," *J. Acoust. Soc. Am.*, vol. 33, 1961, p. 835(A). DOI: 10.1121/1.1936801 2

[7] J. N. Holmes, I. G. Mattingly, and J. N. Shearme, "Speech synthesis by rule," *Language and Speech*, vol. 7, no. 3, 1964, pp. 127-143. 2

[8] G. Fant, *Acoustic Theory of Speech Production.* The Hague, Neitherlands: Mouton, 1970. 3

[9] M. Beutnagel, A. Conkie A., J. Schroeter, Y. Stylianou, and A. Syrdal, "The AT&T next-gen TTS system," *Joint Meeting of ASA, EAA and DAGA*, Berlin, Germany, 1999. 7

[10] A. Black and P. Taylor, "CHATR: A generic speech synthesis system," *Proceedings of the 15th International Conference on Computational Linguistics*, Kyoto, Japan, 1994, pp. 983-986. DOI: 10.3115/991250.991307 7

[11] A. Black and P. Taylor, "The festival speech synthesis system: System documentation," Tech. Rep. HCHC/TR-83, University of Edinburgh, Edinburgh, UK, 1997. 7

[12] W. N. Campbell, "CHATR: A high-definition speech re-sequencing system," *Proc. 3rd ASA/ASJ Joint Meeting*, 1996, pp. 1223-1228. 7

[13] A. Conkie, "Robust unit selection system for speech synthesis," *Joint Meeting of ASA, EAA and DAGA*, Berlin, Germany, March 15 - 19, 1999. DOI: 10.1121/1.425343 7

[14] Y. Stylianou, "Applying the harmonic plus noise model in concatenative speech synthesis," *IEEE Transactions on Speech and Audio Processing*, vol. 9, no. 1, 2001, pp. 21-29. DOI: 10.1109/89.890068 7

[15] Y. Stylianou, "Removing phase mismatches in concatenative speech synthesis," 3rd *ESCA/COCOSDA Workshop on Speech Synthesis*, Jenolan Caves, Australia, Nov. 1998. DOI: 10.1109/89.905997 8

[16] Y. Stylianou, "Concatenative speech synthesis using a harmonic plus noise model," 3rd *ESCA/COCOSDA Workshop on Speech Synthesis*, Jenolan Caves, Australia, Nov. 1998. 8

[17] G. Fant, "Acoustic analysis and systhesis of speech with applications to Swedish," *Ericsson Technics*, vol. 15, 1959, pp. 1615-1626. 9, 53

[18] W. Lawrence, "The synthesis of speech from signals which have a low information rate," in *Communication Theory*, Jackson W. Ed. London, UK: Butterworths Science Publication, 1953, pp. 460-469. 9

[19] J. Anthony and W. Lawrence, "A resonance analogue speech synthesizer," *Proc. 4th Int. Cong. Acoust.*, Copenhagen, Paper G43, 1962, pp. 1-2. 9

[20] J. N. Holmes, "The influence of glottal waveform on the naturalness of speech from a parallel-formant synthesizer," *IEEE Trans. Audio, Elctrroacoust.*, vol. 21, 1973, pp. 298-305. DOI: 10.1109/TAU.1973.1162466 9, 10

[21] W. J. Holmes, J. N. Holmes, and M. W. Judd, "Extension of the band-width of the JSRU parallel-formant synthesizer for high quality synthesis of male and female speech," *Proc. IEEE Int. Conf. on Acoust., Speech, Signal Processing*, vol. 1, 1990, pp. 313-316. DOI: 10.1109/ICASSP.1990.115654 10

[22] J. N. Holmes, "Formant synthesizers: Cascade or parallel," *Speech Communication*, vol. 2, 1983, pp. 251-273. DOI: 10.1016/0167-6393(83)90044-4 10

[23] D. H. Klatt, "Software for a cascade/parallel formant synthesizer," *J. Acoust. Soc. Am.*, vol. 67, 1980, pp. 971-995. DOI: 10.1121/1.383940 10

[24] G. Fant, Q. Lin, and C. Gobl, "Notes on glottal flow interaction," *Speech Transmission Laboratory - Quarterly Progress and Status Report*, Speech Transmission Laboratory, Royal Institute of Technology, Stockholm, vol. 2, 1985, pp. 21-45. 10, 74

[25] J. L. Flanagan, K. Ishizaka, and K. L. Shipley, "Synthesis of speech from a dynamic model of the vocal cords and vocal tract," *Bell Sys. Tech. J.*, vol. 45, no. 3, 1975, pp. 485-506. 11

[26] S. Maeda, "A digital simulation method of the vocal-tract system," *Speech Communication*, vol. 1, 1982, pp. 199-229. DOI: 10.1016/0167-6393(82)90017-6 11

[27] J. L. Kelley and C. C. Lochbaum, "Speech synthesis," *Proc. 4th Int. Congr. Acoust.,* Copenhagen, Paper G-42, 1962, pp. 1-4. 11

[28] J. Liljencrants, "Speech synthesis with a reflection-type analog," Ph.D. dissertation, Royal Institute of Technology, Stockholm, Sweden, 1985.

[29] P. Meyer, R. Wilhelms, and H. W. A. Strube, "Quasiarticulatory speech synthesizer for German language running in real time," *J. Acoust. Soc. Am.,* vol. 86, no. 2, 1989, pp. 523-539. DOI: 10.1121/1.398232 11

[30] M. M. Sondhi and J. Schroeter, "A hybrid time-frequency domain articulatory speech synthesizer," *IEEE Trans. Acoust., Speech, Signal Processing,* vol. 35, no. 7, 1987, pp. 955-967. DOI: 10.1109/TASSP.1987.1165240 11

[31] S. E. Levinson and C. E. Schmidt, "Adaptive computation of articulatory parameters from the speech signal," *J. Acoust. Soc. Am.,* vol. 74, no. 4, Oct. 1983, pp. 1145-1154. DOI: 10.1121/1.390038 11

[32] M. A. Hasegawa-Johnson and J. S. Cha, "CTMRedit: A Matlab-based tool for viewing, editing and three-dimensional reconstruction of MR and CT images," *Proc. BMES/EMBS,* Atlanta, FL, 1999. 11, 20, 45, 46, 80

[33] M. G. Rahim and C. C. Goodyear, "Parameter estimation for spectral matching in articulatory synthesis," *Collq. Spectral Est. Tech. Speech Processing,* London, UK: IEE Press, 1989. 11

[34] M. G. Rahim and C. C. Goodyear, "Articulatory speech with the aid of a neural net," *Proc. IEEE Intl. COnf. on Acous., Speech, Signal Processing,* vol. 1, 1989, pp. 227-230. 12

[35] M. G. Rahim and C. C. Goodyear, "Estimation of vocal tract filter parameters using a neural net," *Speech Communication,* vol. 9, 1990, pp. 49-55. DOI: 10.1016/0167-6393(90)90045-B 12

[36] M. G. Rahim, *Artificial Neural Networks for Speech Analysis/Synthesis.* London, UK: Chapman & Hall, 1994. 12

[37] L. R. Rabiner and R. W. Schafer, *Digital Processing of Speech Signals.* Englewood Cliffs, NJ: Prentice Hall, 1978. 12

[38] J. D. Markel and H. Gray, "On autocorrelation equations as applied to speech analysis," *IEEE Trans. Audio Electroacoust.,* vol. 20, 1973, pp. 69-79. DOI: 10.1109/TAU.1973.1162440 13, 14

[39] B. S. Atal and L. Hanauer, "Speech analysis and synthesis by linear prediction of the speech wave," *J. Acoust. Soc. Am.,* vol. 50, 1971, pp. 637-655. DOI: 10.1121/1.1912679 13

[40] D. Sinder, "Speech synthesis using an aeroacoustic model," Technical Report CAIP-TR-236, CAIP center of Rutgers University, 1999. 13

[41] M. S. Howe, *Acoustics of Fluid-Structure Interactions.* Cambridge, UK: Cambridge University Press, 1998. DOI: 10.1017/CBO9780511662898 14, 17

[42] K. Ishizaka and J. L. Flanagan, "Synthesis of voiced sounds from a two-mass model of the vocal cords," *Bell Sys. Tech. J.*, vol. 51, no. 6, 1972, pp. 1233-1268. 14

[43] I. G. Currie, *Fundamental Mechanics of Fluids.* New York: McGraw-Hill Inc., 1974. 15, 54, 60

[44] M. H. Krane, "Aeroacoustic production of unvoiced speech sounds," submitted to Journal of the Acoustical Society of America. DOI: 10.1121/1.1862251 17

[45] C. H. Coker, "A model of articulatory dynamics and control," *Proc. IEEE*, vol. 64, 1973, pp. 452-460. DOI: 10.1109/PROC.1976.10154 19, 20

[46] P. Mermelstein, "Articulatory model for the study of speech production," *Journal of the Acoustical Society of America*, vol. 53, 1973, pp. 1070-1082. DOI: 10.1121/1.1913427 20, 51, 53

[47] P. Rubin, T. Baser, and P. Mermelstein, "An articulatory synthesizer for perceptual research," *J. Acoust. Soc. Am.*, vol. 70, no. 2, 1981, pp. 321-328. DOI: 10.1121/1.386780 23, 24, 51, 53

[48] E. L. Saltzman and K. G. Munhall, "A dynamical approach to gestural patterning in speech production," *Haskin Laboratories Status Rerport on Speech Production*, SR-99/100, 1989, pp. 38-68. DOI: 10.1207/s15326969eco0104_2 25

[49] E. L. Saltzman, "The task dynamic model in speecg production," in *Speech motor control and shuttering*, Peters H., Hulstijn W. and Starkweather C. W., Eds. New York: Elsevier Science Publishers, 1991, pp. 37-52. 25, 26, 27, 51, 53

[50] P. Rubin, E. Saltzman, L. Goldstein, R. McGowan, M. Tiede, and C. Browman, "CASY and extensions to the task-dynamic model," *1st ECSA Tutorial and Research Workshop on Speech production Modeling - 4th Speech Production Seminar*, 1996, pp. 125-128. 25

[51] N. Chomsky and M. Halle, *The Sound Pattern of English.* New York: Harper & Row, 1968. 25

[52] W. L. Nelson, "Physical principles for economies of skilled movements," *Journal of Biological Cybernetics*, vol. 46, 1983, pp. 135-147. DOI: 10.1007/BF00339982 26

[53] M. Unser and I. Daubechies, "On the approximation power of convolution-based least squares versus interpolation," *IEEE Trans. on Signal Processing*, vol. 45, no. 7, 1997, pp. 1697-1711. DOI: 10.1109/78.599940 29

[54] T. Blu and M. Unser, "Quantitative Fourier analysis of approximation techniques: Part I – interpolators and projectors," *IEEE Trans. on Signal Processing*, vol. 47, no. 10, 1999, pp. 2783-2795. DOI: 10.1109/78.790659 34, 35, 38, 39, 40, 41, 42, 43, 46

[55] G. Strang and G. Fix, "A Fourier analysis of the finite element variational method," in *Constructive Aspect of Functional Analysis*. Rome, Italy: Cremonese, 1971, pp. 796-830. 34

[56] S. Maeda, "An articulatory model of the tongue based on a statistical analysis," in *Speech Communication Papers*, Wolf J. J., and Klatt D. H. Eds. New York: Acoustical Society of America, 1979, pp. 67-70. 38

[57] K. Shirai and M. Honda, "An articulatory model and the estimation of articulatory parameters by nonlinear regression method," *Electron. Comm.*, Japan 59-A, 1976, pp. 35-43. 51, 53

[58] V. N. Sorokin, "Determination of vocal tract shape for vowels," *Speech Communication*, vol. 11, 1992, pp. 71-85. DOI: 10.1016/0167-6393(92)90064-E 51, 53

[59] N. Nguyen, P. Hoole, and A. Marchal, "Regenerating the spectral shape of [*s*] and [*ʃ*] from a limited set of articulatory parameters," *Journal of the Acoustical Society of America*, vol. 96, 1994, pp. 33-39. DOI: 10.1121/1.411435 51, 53

[60] K. N. Stevens, "On the quantal nature of speech," *Journal of Phonetics*, vol. 17, 1989, pp. 3-45. 51, 53

[61] R. Harshman, P. Ladefoged, and L. Goldstein, "Factor analysis of tongue shapes," *Journal of the Acoustical Society of America*, vol. 62, 1977, pp. 693-707. DOI: 10.1121/1.381581 51, 53

[62] V. N. Sorokin, A. S. Leonov, and A. V. Trushkin, "Estimation of stability and accuracy of inverse problem solution for the vocal tract," *Speech Communication*, vol. 30, 2000, pp. 55-74. DOI: 10.1016/S0167-6393(99)00031-X 51, 53

[63] J. N. Holmes, "An investigation of the volume velocity waveform at the larynx during speech by means of an inverse filter," *Proc. Speech Communication Seminar*, Royal Institute of Technology, Stockholm, Sweden, vol. 1, 1962, pp. B4. 52

[64] M. Rothenberg, "A new inverse-filtering technique for deriving the glottal air flow waveform during voicing," *J. Acoust. Soc. Am.*, vol. 53, no. 6, 1973, pp. 1632-1654. DOI: 10.1121/1.1913513 59

[65] A. E. Rosenberg, "Effects of pulse shape on the quality of natural vowels," *J. Acoust. Soc. Am.*, vol. 49, no. 2, 1973, pp. 583-591. DOI: 10.1121/1.1912389 59

[66] I. R. Titze, "Synthesis of sung vowels using a time-domain approach," in *Transcripts of the 11th Symp.: Care of the Prof. Voice*, V. L. Lawrence Ed. New York: The Voice Foundation, 1982, pp. 90-98. 59

[67] Y. Koizumi, S. Taniguchi, and S. Hiromitsu, "Two-mass models of the vocal cords for natural sounding voice synthesis," *J. Acoust. Soc. Am.*, vol. 82, 1987, pp. 1179-1192. DOI: 10.1121/1.395254 59, 60

[68] B. H. Story and I. R. Titze, "Voiced simulation with a body-cover model of the vocal folds," *J. Acoust. Soc. Am.*, vol. 97, 1995, pp. 1249-1260. DOI: 10.1121/1.412234 61

[69] D. A. Berry, H. Herzel, I. R. Titze, and K. Krischer, "Interpretation of biomechanical simulations of normal and chaotic vocal fold oscillations with empirical eigenfunctions," *J. Acoust. Soc. Am.*, vol. 95, 1994, pp. 3595-3604. DOI: 10.1121/1.409875 61

[70] D. A. Berry and I. R. Titze, "Normal modes in a contunuum model of vocal fold tissues," *J. Acoust. Soc. Am.*, vol. 100, 1996, pp. 3345-3354. DOI: 10.1121/1.416975 61

[71] G. Fant and T. V. Ananthapadmanabha, "Speech production," *STL-QPSR*, Speech Transmission Laboatory, Royal Institute of Technology, Stockholm, vol. 2, 1982, pp. 1-17. 61

[72] G. Fant and Q. Lin, "Glottal source - vocal tractacoustic interaction," *STL-QPSR*, Speech Transmission Laboatory, Royal Institute of Technology, Stockholm, vol. 1, 1987, pp. 13-27. DOI: 10.1121/1.2024357 74

[73] M. S. Howe, "The genaration of sound by aerodynamic sources in an inhomogeneous steady flow," *J. Fluid Mechanics*, vol. 67, part 3, 1975, pp. 597-610. DOI: 10.1017/S0022112075000493 74

[74] P. J. Brockwell and R. A. Davis, *Time Series: Theory and Methods.* New York: Springer-Verlag Inc., 2nd edition, 1991. DOI: 10.1007/978-1-4419-0320-4

[75] P. J. Brockwell and R. A. Davis, *Introduction to Time Series and Forecasting.* New York: Springer-Verlag Inc., 1996.

[76] M. P. Clements and D. F. Hendry, *Forcasting Non-stationary Economic Time Series.* Cambridge: The MIT Press, 1999.

[77] H. Tong, *Non-linear Time Series.* New York: Oxford University Press, 1990.

[78] L. R. Rabiner and S. E. Levinson, "Isolated and connected word recognition - theory and selected applications," *IEEE Trans. Communications*, vol. 29, no. 5, 1981, pp. 621-659. DOI: 10.1109/TCOM.1981.1095031

[79] C.H. Coker, "A model of articulatory dynamics and control," *Proceedings of the IEEE*, Vol. 64, No. 4, pp. 452–460, Apr. 1976. 93

[80] S.E. Levinson and C.E. Schmidt, "Adaptive computation of articulatory parameters from the speech signal," *J. Acoust. Soc. Am.*, 74(4), pp. 1145–1154, Oct. 1983.

[81] E.L. Saltzman and K.G. Munfall, "A dynamical approach to gestural patterning in speech production," *Haskind Laboratories Status Report on Speech Production*, SR-99/100, pp. 38–68, 1989.

[82] N. Chomsky and M. Halle, *The Sound Pattern of English*, New York: Harper & Row, 1968.

[83] W.L. Nelson, "Physical principles for economies of skilled movements," *Journal of Biological Cybernetics*, vol. 46, pp. 135–147, 1983.

[84] G. Thimm and J. Luettin, "Extraction of articulators in X-ray image sequences," *Proceedings of Eurospeech*, Hungary, Sept. 1999.

[85] X. Pelorson, A. Hirschberg, R.R. Van Hassel, A.P.J. Wijnands, Y. Auregan, "Theoretical and experimental study of quasi-steady flow separation within the glottis during phonation: Application to a two-mass model," *J. Acoust. Soc. Am.*, 1994 (96), pp. 3416–3431.

54

Authors' Biographies

STEPHEN E. LEVINSON

Stephen E. Levinson was born in New York City on September 27, 1944. He received a B.A. degree in Engineering Sciences from Harvard in 1966, and M.S. and Ph.D. degrees in Electrical Engineering from the University of Rhode Island, Kingston, Rhode Island in 1972 and 1974, respectively. From 1966–1969 he was a design engineer at Electric Boat Division of General Dynamics in Groton, Connecticut. From 1974–1976 he held a J. Willard Gibbs Instructorship in Computer Science at Yale University. In 1976, he joined the technical staff of Bell Laboratories in Murray Hill, NJ where he conducted research in the areas of speech recognition and understanding. In 1979, he was a visiting researcher at the NTT Musashino Electrical Communication Laboratory in Tokyo, Japan. He held a visiting fellowship in the Engineering Department at Cambridge University in 1984, and in 1990 he became head of the Linguistics Research Department at AT&T Bell Laboratories where he directed research in Speech Synthesis, Speech Recognition, and Spoken Language Translation. He joined the Department of Electrical and Computer Engineering of the University of Illinois at Urbana-Champaign in 1997, where he teaches courses in Speech and Language Processing and leads research projects in speech synthesis and automatic language acquisition. He is also a full-time faculty member of the Beckman Institute for Advanced Science and Technology where he serves as the head of the Artificial Intelligence group. Dr. Levinson is a member of the Association for Computing Machinery, a fellow of the Institute of Electrical and Electronic Engineers, and a fellow of the Acoustical Society of America. He is a founding editor of the journal *Computer Speech and Language* and a former member and chair of the Industrial Advisory Board of the CAIP Center at Rutgers University. He is the author of more than 100 technical papers and holds 7 patents. His book, published in 2005 by John Wiley and Sons, Ltd., is entitled *Mathematical Models for Speech Technology*.

DON W. DAVIS, JR.

Donald W. Davis, Jr. received B. S., M. S., and Ph. D. degrees in Aeronautical Engineering from Purdue University in 1970, 1975, and 1981, respectively. Currently, he is a Staff Engineer at Electric Boat Corporation where he works in the area of computational fluid dynamics (CFD). His research interests include fluid mechanics, heat transfer, computational methods, and turbulence modeling. He is also involved in applying advanced CFD tools to large, complex, industrially relevant turbulent flow problems.

SCOT A. SLIMON

Scot A. Slimon received a B.S. in Marine Engineering Systems from the United States Merchant Marine Academy, an M.S. in Mechanical Engineering from the Rensselaer Polytechnic Institute, and a Ph.D. in Mechanical Engineering from the University of Connecticut. Currently, he is a Principal Engineer at Electric Boat Corporation, where he is responsible for the development and application of a computational fluid dynamics solver. His current research involves preconditioning techniques, multiphase flow, hybrid turbulence modeling, and flow induced sound at low Mach numbers. He has applied this research to a number of large-scale external and internal flow problems supporting major Navy submarine platforms.

Printed in the United States
by Baker & Taylor Publisher Services